Lewis Sherman

Therapeutics and materia medica for the use of families and physicians

Second Edition

Lewis Sherman

Therapeutics and materia medica for the use of families and physicians
Second Edition

ISBN/EAN: 9783337872564

Printed in Europe, USA, Canada, Australia, Japan

Cover: Foto ©berggeist007 / pixelio.de

More available books at **www.hansebooks.com**

— AND —

MATERIA MEDICA

FOR THE USE OF

FAMILIES AND PHYSICIANS.

SECOND EDITION.

BY

LEWIS SHERMAN, A. M., M. D.

PREFACE.

The object of this book is, to instruct intelligent, non-professional persons in the nature, symptoms, course, prevention and treatment of the most common forms of disease.

It is not intended to usurp the place of the educated and experienced family physician, but rather to aid him by diffusing knowledge among his patients. Domestic medicine is a necessity. It may be reformed, but it cannot be abolished. Herb Teas, Tonic Bitters, Liver Pills, Soothing Syrups and Pain Killers are to be found in every household, and are often administered before the physician is called. The choice lies between the unguided, or misguided, use of powerful agents, and the careful and better directed use of milder means of cure.

While, by its freedom from unnecessary technicalities, the work is adapted to the comprehension of the non-professional, it is hoped that it is sufficiently accurate to be useful to the practitioner of Medicine.

The introduction of some of the most valuable of the new remedies, will find favor with the profession. Gelsemium, Cimicifuga, Hamamelis, Hydrastis, Podophyllum, and other indigenous American remedies, form a prominent part of the armamentarium of the successful physician, yet these remedies are scarcely recognized in other works having the same scope as this. Such medicines as Chininum arsenicosum, Eucalyptus and Physostigma, which have been more recently brought to the notice of the medical profession, but which have not yet come into general use, are occasionally recommended.

In conclusion, we trust that the prominence given to auxiliary means of cure, such as the influences of food, drink, air, exercise, rest and bathing, and the use of disinfectants and local applications, will not be regarded as an unimportant feature of the work.

LEWIS SHERMAN.

MILWAUKEE, WIS.

INTRODUCTION.

HYGIENE.

Pure air, pure water, nutritious food, sun-light, exercise, rest and, in this climate, clothing are essentials to health.

PURE AIR.—This means out-door air, whether in the day or night. The popular prejudice against "night air" is unfounded. If strong drafts and extreme changes of temperature can be avoided, it is better to have the windows of sleeping-rooms open at night. It is safer, in cold weather, to have the opening as low as possible, to avoid unnecessary cooling of the air. The principle of "diffusion of gases" will insure the necessary supply of oxygen. Even asthmatics, who are very sensitive to atmospheric changes, sleep better in a well ventilated room than in the most carefully heated, closed room.

PURE WATER.—This means water which is free from poisonous minerals and from decaying animal and vegetable substances. The best drinking water contains oxygen gas and some mineral and saline ingredients in solution. Pure, soft spring-water is to be preferred for drinking and cooking purposes. Where this cannot be obtained, well-water, in the country, and rain-water, filtered through sand and charcoal, in the city, are next

2

in excellence. As a rule, water is more or less contaminated by standing in lead pipes. The amount of contamination varies with the composition of the water; the softer the water the more likely it is to dissolve the lead of pipes through which it passes. The taste will detect the poisonous substance when a large quantity is present; but there is no doubt that water containing lead, in much smaller proportions than can be revealed by the taste, is deleterious to health. Impure water may be rendered comparatively harmless by prolonged boiling. After boiling, it should be allowed to stand exposed to the fresh air or be shaken in a large vessel in order to render it palatable by the absorbtion of oxygen. Ice may be regarded as pure if it is perfectly transparent; otherwise it may be as poisonous as the water of the ponds, rivers and canals, from which it is taken.

NUTRITIOUS FOOD.—This means food which contains all of the proximate elements of the body, in the proportions to supply the processes of growth and repair. Food, to be nutritious, must be digestible, that is, soluble in the fluids of the alimentary canal.

Wheat is the most nutritious grain known. The outer husk is indigestible, therefore innutritious, and should be rejected by all except those of constipated habit, who lead a sedentary life. The inner coat is essential to the full nutritive value of the grain. This is usually removed with the outer coat. The kernel contains the starchy portion of the grain which, though important, is deficient in the nitrogenous or tissue-repairing elements.

Bread should be light, but care should be taken not

to use too much yeast nor to let the dough rise too long; for over-raising injures the nutritious properties as well as the flavor of bread. Stale bread is more easily digested than fresh, and should be preferred by invalids.

Indian Corn is a less valuable grain than wheat. It contains a smaller proportion of nutriment and is less easily digested. Corn cakes and corn-meal mush are beneficial in constipated habit of the bowels.

Oats make very nutritious meal. Oat-meal mush and milk makes almost a complete diet.

Rice is a mild and easily digested grain, useful to invalids and convalescents, especially in irritable states of the alimentary canal.

Iceland Moss is very nutritious and easily digested. It it is useful in bronchial affections.

Malt—partially fermented barley—contains a peculiar active nitrogenous principle called *diastase*, which has the power of converting starch into dextrine and sugar, thus rendering it soluble. An infusion of malt is made by boiling, for ten minutes, four tablespoonfuls of ground malt in a pint of water. It is an agreeable, refreshing and nutritive drink, which is useful in Dyspepsia and diseases accompanied by weak digestion.

Potatoes are nutritious, but oily and nitrogenous foods must be eaten with them. One of the best methods of cooking potatoes is roasting. If steamed or boiled they should be cooked before the skins are removed, and a little common table salt should be added to the water to prevent the escape of the nutritious and palatable natural salts.

Pastry should be eaten sparingly by the healthy and not at all by those of weak digestion.

Animal food,—whether milk, eggs or the parts of animal bodies,—contains the most nutriment for its bulk and weight of all classes of food.

Milk is perfect food for the young and, with the addition of starchy and fatty foods, also for the adult. Human milk differs from cow's milk in containing more milk-sugar and more water with less cheese and butter. If cow's milk be used for infants, as a substitute for mother's milk, it should receive the addition of one-third its bulk of a hot solution of sugar of milk, containing half an ounce, or a small tablespoonful, to the pint. The milk should not be boiled. When there is a tendency to acidity of the stomach, the addition of lime-water or carbonate of soda to milk renders it more easily retained and digested. One part of transparent lime-water to two or three parts of milk, or fifteen grains of carbonate of soda to the quart, is the proper proportion.

Eggs are exceedingly rich in nutritive properties. They are often cooked so as to render them difficult of digestion. They are most easily digested in the raw state, but there is no objection to hardening the white.

Beef, although not the most easily digested of animal meats, is the most generally satisfactory, on account of its peculiar flavor and its better hunger-appeasing qualities.

Mutton is in some cases to be preferred to beef on account of its greater softness of texture and consequent ease of digestion.

Chicken is a valuable meat for invalids and con-valescents.

Turkey, *Goose* and *Duck* are too strong in flavor and too tough for weak stomachs.

Squabs, (young pigeons), are tender and delicate, and may be eaten by convalescents.

Wild birds are all edible, and some afford rich delica-cies for the sick.

The flesh of birds is said by the chemists to be as nutritious as that of mammals; but it is not as satisfying to the hungry and therefore not so generally eaten as the flesh of the cow and sheep. The flesh of game is tougher than that of domestic fowls, and for this reason requires to lie several days after the birds are killed.

The *cooking* of meat is almost as important a matter as its selection. If meat is to be boiled, it should be put in a vessel of boiling water in order to harden the out-side and prevent the egress of the soluble parts. The remainder of the process of cooking should be at a temperature much below the boiling point,—about 160° to 180° F. The fibers are thus separated and the meat is made soft and retains its flavor and nutritive proper-ties. If meat is to be baked, it should be put, at first, into a hot oven, and, when the outside is hardened, the heat of the fire should be diminished. Meat thus cooked retains its flavor and the greater portion of the volatile nutritious parts which would otherwise escape. Raw beef, pounded to a jelly, with the addition of salt, and, in some cases, of sugar or black pepper, is a highly nutritious and stimulating food which may be adminis-tered to patients who are greatly reduced in strength.

If *soup* or *broth* is desired, the meat should be put into cold water, heated slowly and cooked a long time.

Beef Tea is a valuable dish in cases of debility or prostration and in some forms of Dyspepsia. Although far less nutritive than the beef itself, it contains the flavor and stimulant properties of the meat, and creates a desire for other food. The same may be said of *mutton broth* and *chicken broth.*

Pork, though highly nutritious, is not so good a food as beef and mutton. It disagrees with many stomachs. On account of the danger from the pork-worm, it should never be eaten raw.

Fish is very rich in nutriment and easy of digestion; though like poultry it does not satisfy the demands of hunger as well as other flesh. It might well be used to a greater extent than it is at the present time.

Oysters from September to April, i. e., except in the *r-less* months, are a valuable food. They are most easily digested in the raw state.

Fruits, as apples, pears, grapes, berries, currants, cherries, plums, peaches, oranges, bananas, etc., in the ripe state, may be eaten raw. They are the principal subsistence of the inhabitants of hot regions, and are too much neglected in the hot seasons of the temperate latitudes. They promote the secretions and prevent bilious disorders.

Tea may be used by the feeble and the old as a habitual drink, and by the young and robust during fatigue. Persons suffering from chronic Constipation should drink it sparingly or not at all. Tea, for ordinary use, should be steeped, from three to five minutes. Prolonged boiling extracts the *Tannic Acid* of the leaves and makes the infusion bitter and astringent.

Coffee has stimulant properties similar to those of Tea, though its flavor is different. The aroma is developed by the roasting process, and much skill is required to prepare a palatable drink.

Chocolate is not a stimulant, but contains considerable oil and sugar. It can be used as a beverage, with impunity, by most persons. The large quantity of oil which it contains makes it objectionable to some dyspeptics.

Time of Meals.—As a general rule, the American custom of eating three meals, daily, and making the middle one the heartiest, is most conducive to health.

Food should not be bolted or washed down with drinks; but should be brought to a pulp before it is swallowed.

Sun-Light.—The influence of sun-light, upon the animal spirits and vigor, is well known, but too little insisted upon. The exposure of the naked person to the direct rays of the sun has been found a valuable therapeutic means. The interior of houses should be exposed as much as possible to strong sun-light, to kill the fungi and other low forms of life which infest dark and damp places and breed miasm.

Exercise.—Daily exercise in the open air is an absolute essential to robust health. Walking is the best mode of exercise for persons of in-door habits, because it is thorough, mild and entertaining. From one to six miles is the least distance allowable to those of ordinary health, who take no physical exercise. The proper time is from one to four hours after a meal.

REST.—The necessary amount of sleep is usually indicated by the feelings. Young infants require from twelve to sixteen hours, children, about ten hours, adults, eight hours. Individual peculiarities cause slight variations from these rules, but any considerable variation denotes a departure from health. Regular habits of sleep have much to do with longevity.

CLOTHING.—Clothing should be tempered to the weather and the amount of exercise taken. The feet, legs and back are the parts most sensitive to cold, and require the greatest amount of covering. The senseless practice of dressing children in such a manner as to leave the lower limbs half protected, cannot be too severely condemned. Mothers often sacrifice the health and even the lives of their children to gratify their own vanity. The proper temperature of rooms depends much upon the occupation of the inmates. For active exercise, no artificial heat may be required; for refreshing sleep 50° to 60° F. is sufficiently warm; while for sedentary employments, which do not bring the blood into active circulation, the temperature of 68° F. is preferable.

Bathing is one of the luxuries of life and not one of the essentials to health. The cold plunge may be safely indulged in, by those of healthy circulation. It should not last over five minutes,—never long enough to produce a sense of chilliness, which continues after the body is out of the water. The tepid bath and the sponge bath are exceedingly grateful and often beneficial to those suffering from fever. The warm bath is best for cleansing purposes and to promote the secretions of the skin, in diseases which require the eliminative function of this organ.

MEDICINES.

Medicines are poisonous substances, which, when introduced into a diseased animal system, tend to restore healthy action. The idea that medicines are essentially good and work in harmony with the healthy organism, is as mischievous as it is erroneous. The glaring adver. tisements of "Bitters," "Tonics," "Blood-Purifiers" and other vile nostrums which flood the land, are demanded and paid for by a class of people who are grossly ignorant of physiological and hygienic laws. The medicines which cultured and experienced physicians of every school find most useful in the cure of disease, are those which produce in the healthy organism the most poisonous effects; for example, Arsenic, Morphine, Strychnine, Quinine, Aconite, Belladonna, Mercury and Chloroform are all deadly poisons. Poisons become curative only when they are adapted to the diseased organism and reduced in dose. The law of adaptation is expressed in the Latin formula, "*Similia similibus curantur*,"—freely translated, "Like cures like." Quinine in large doses will produce, and in smaller doses will cure Intermittent Fever. Ipecacuanha in large doses will produce, and in small doses will cure spasmodic Asthma and vomiting. Tartar emetic in large doses will produce, and in small doses will cure Inflammation of the Lungs. Nux vomica, in large doses will produce, and in small doses will cure Constipation. Belladonna in large doses will produce, and in small doses will cure Headache and Sore-Throat.

This law of cure probably bears some relation to the law of drug action, namely, that poisons have a primary and a secondary action, opposite in character. If one takes a cathartic, the primary effect is a medicinal Diarrhœa, and the secondary effect, of longer duration, is Constipation. Opium or Morphine in a large dose produces, for its primary effect, stupor and sleep and, for its secondary effect, nervousness and sleeplessness, of much longer duration. When small doses are given, the primary effect is less marked and the secondary effect comes on sooner.

The law, "Like cures like," established by observation and experiment, enables us to select with precision the remedies for the cure of disease. Hundreds of drugs have been given to the healthy to determine their effects. The results of these experiments, called "provings," are recorded and published under the title of "Materia Medica." The selection of the remedy consists in finding a drug, of which the poisonous effects resemble the symptoms of the disease.

STRENGTH OF MEDICINES.—The strength of the medicines recommended in this book, if not stated in the text, is indicated in the "List of Medicines" given just before the section on Materia Medica.

DOSE.—The dose of the liquid preparations, is for an adult, uniformly, one drop in a teaspoonful of water, unless otherwise stated. This dose may be conveniently

administered by dropping from ten to fifteen drops into
a tumbler and adding as many teaspoonfuls of water.
The dose of the solution will then be one teaspoonful.
By holding the cork against the lip of the vial in the
manner represented in the cut, the process of counting
drops is rendered easy.

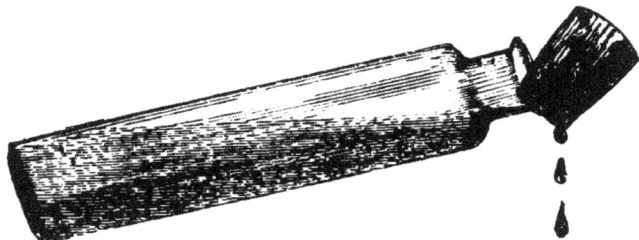

The dose for a child may be reduced by putting in from
two to six teaspoonfuls of water to the drop.

The medicine glass should be perfectly clean, and in
most cases it is better not to leave a spoon in the glass,
but to wipe it dry, each time, after using. The pellets
may be taken directly into the mouth and swallowed,
with or without chewing. The dose for an adult is six;
for the youngest infant, one; for children, from two to
five. The dose of the trituration is, for an adult, one
grain, i. e., the weight of a large grain of wheat. The
dose may be reduced for infants and children by dissolving it in water and giving in divided doses.

REPETITION.—The repetitions of the dose should be, in
acute diseases, from half an hour to two hours apart; in
chronic diseases, from one to three times a day.

Alternation of remedies is sometimes practiced in cases
in which one remedy cannot be found which will cover
all the symptoms of the case.

Mixing medicines should be avoided as much as possible, because the effects of mixed medicines may not be the same as those of the constituents, used separately.

Cathartics, emetics, blisters, etc., should never be administered without the advice of a competent educated physician.

PART I.

THERAPEUTICS.

APOPLEXY.

SYMPTOMS.—This disease is characterized by the sudden loss of the senses and of the power of motion and thought, caused by a sudden pressure upon the brain, originating within the cranium. In most cases there is an escape of blood from the blood-vessels into the substance of the brain. In rare cases there is only a congested condition of the blood-vessels of the brain.

In Apoplexy from *congestion* there may be some premonitory symptoms, lasting for a few hours or minutes, such as, confusion of thought, drowsiness, inactivity, dimness of sight, or the appearance of floating specks in the field of vision, dizziness, rumbling noises in the head and headache. The attack seems to be precipitated by some extra exertion, as lifting a heavy weight, pulling on the boots, blowing the nose, or straining at stool. The attack then comes suddenly, and there is partial or complete loss of consciousness and sensation, deafness, blindness and paralysis. In the course of a few minutes, these severe symptoms abate and the patient gradually recovers the mental and physical powers.

In apoplexy, from the *bursting of a blood-vessel*, there are generally no premonitory symptoms, and the loss of the senses and the paralysis are more complete. The paralysis may be on one side or on both sides of the body. Recovery from this form is not so rapid nor so complete, and the patient rarely survives the third attack.

Causes.—The use of Alcohol, Opium or Tobacco, intemperance in eating, overheating of the brain, blows upon the head. Short-necked, florid and inactive people, are most subject to Apoplexy, though thin and long-necked, nervous people, are sometimes attacked. In the latter class it is generally hereditary.

Treatment.—For the congestive symptoms, give *Bell*. and *Arn*. in alternation. If there is indigestion and constipation of the bowels, give *Nux*. The patient should be kept as quiet as possible; the head cool and slightly raised. The diet should be very light food, till danger of an immediate relapse is passed. Milk, bread, light vegetables and fish may be allowed, but the patient should never return to the use of a full animal diet, or indulge in alcoholic stimulants; and all the causes above referred to should be cautiously avoided.

APPETITE.
IRREGULARITIES OF.

Treatment.—*Diminution or loss of appetite*, in acute disorders, accompanied by fever, does not demand special treatment. When it is persistent and not accompanied by other marked symptoms, there is generally some fault in the digestive process. The administration of *Nux*., *Arsen*. or *Puls*. will generally improve the digestion and effect a cure.

The indications for *Nux.* are constipation, dull headache or drowsiness.

Arsen. is preferable if there are sour eructations, fullness of the stomach and a tendency to looseness of the bowels.

Puls. is appropriate in nervous temperaments, and when there are shifting pains and other abnormal sensations in the stomach.

Voracious appetite may also be a symptom of indigestion. In children it is often caused by the presence of worms in the alimentary canal. A dose of *Santonine*, third decimal trituration, given three times a day, before eating, will in most cases be found curative. If the appetite is *capricious*, *Puls.*, given in the same manner, is the appropriate remedy. The habit of eating between meals is often the cause of irregular or depraved appetite. The hours for meals should be regular, and, except in rare instances, three meals a day is enough.

ASTHMA.

This is a constitutional disease, which manifests itself in paroxysms of difficult breathing. The breathing is accompanied by a wheezing sound and a tightness in the chest, which prevents the patient from getting a full breath. The paroxysm usually ends after the expectoration of a quantity of mucus from the bronchial tubes. The attack may come once a year, as in Winter Asthma and Hay or Summer Asthma; once a month, in women who are subject to the disease; at every exposure to cold or severe exercise, in others; and, in bad cases, every night.

Asthma may depend upon disease of the heart, in

which case it is always produced by severe exercise or excitement; or it may depend upon impaired digestion, when the paroxysms usually follow excesses in diet; but the most common cause is Chronic Bronchitis.

TREATMENT.—*Arsen.* will greatly modify the difficult breathing of *heart-disease.* In some cases *Arsen.*, in alternation with *Cactus grandiflorus*, second decimal dilution on pellets, acts like a charm. All violent exercise should be avoided.

In the *dyspeptic* form, *Arsen.* and *Ipec.* are the most useful remedies. All excesses in diet must be most sedulously avoided. The meals should be light. Fish and mutton are preferable to beef and pork, because they remain a shorter time in the stomach.

Asthma from *Chronic Bronchitis*, is one of the most distressing complaints known. Some cases are incurable; but most are benefited, and a few are entirely cured by a change of climate. Patients in the southern and eastern parts of the United States are often benefited by the dry air and high altitude of Minnesota. The Straits of Mackinaw afford immunity to some; the arid plateaus of Texas, to others; and the mild equable climate of Southern California, to others; but the great resort for asthmatics is the elevated mountainous region of Colorado. The journey to Colorado should be slow, and, if the lungs are diseased, some months should be occupied in reaching the altitude of eight to ten thousand feet. A life in the open air is essential to the full beneficial influence of the climate.

The distressing paroxysms of Asthma may be relieved by inhaling the smoke of the dried leaves of *Stramonium*, or by burning paper which has been soaked in a solution of *Nitrate of Potash.* Several other substances have been used, with varying success, as inhalations to relieve the paroxysms. Among these are the leaves of *Belladonna, Silphium laciniatum* and *Silphium terebinthinaceum.* None of these inhalations cure the disease, and all, by continued use, lose their power to modify the severity of the attacks.

Arsen. may be taken with great benefit before and during the paroxysm, especially if there is a thin, watery secretion from the mucous membrane, cold sweat and general prostration of strength.

Tart. emet. is very useful when there is a profuse secretion of mucus which is raised with difficulty.

Kali iod. will relieve if there are symptoms of an acute Catarrh, indicated by sneezing, watering of the eyes and general stuffed condition of the air passages of the head.

If there is no discharge use *Ipecac.*

If the throat seems filled with a very tough tenacious phlegm, which is detached with difficulty, use *Kali bichrom.*

BED-SORES.

Patients suffering from protracted fevers and wasting diseases, such as Paralysis and Consumption, frequently suffer from the pressure of the back or hips upon the bed. The first indications of approaching trouble are, slight tenderness and occasional redness of the skin.

TREATMENT.—Bed-sores are more easily prevented than cured. Patients who are confined to the bed, for a long time, should be frequently bathed, to prevent the

3

irritation of the skin, produced by the perspiration. The bed-linen should be well ironed, and changed often, and care should be taken to prevent the patient from lying on wrinkles. When there is the least irritation of the skin, the parts affected should be washed, daily, with dilute *Alcohol.* If the skin is broken, it should be moistened with a solution of *Tannin* and *Glycerine,*— about one drachm to the ounce. It is well to arrange cushions and pillows in such a manner as to protect the tender parts from pressure.

BOILS.

A boil is a red, shining, pointed, tender and painful tumor, which is filled with pus when mature. There is a disordered condition of the blood, in which the blood corpuscles degenerate and accumulate in the structures of the body where they break down the tissues and escape.

CAUSES.—Insufficient nutrition, imperfect digestion, and bad food and water, are the common causes.

TREATMENT.—The local means are the application of heat and moisture. A warm poultice of flax-seed meal is as good as anything. It should be changed as often as it gets cold. If the poultice is covered with oiled silk and flannel, it will not require so frequent changing.

The constitutional treatment is: for the acute symptoms, *Bell.* and *Hepar Sulph.* To prevent the recurrence, take *Sulphur* two or three times a day. The diet should be simple, but not restricted to vegetable food. Cold bathing and out-door exercise will help to remove the predisposition to boils.

BRONCHITIS.

DEFINITION.—This is essentially an inflammation of
the mucous membrane of the bronchial tubes, which are
the branches of the windpipe.

SYMPTOMS.—Hoarseness, cough, and soreness of the
chest in the region of the breast bone, are the prominent
symptoms of acute Bronchitis. The mucous secretion
is at first arrested; afterward, it becomes profuse and
changes from a watery to a thick, white, yellowish or
greenish discharge. Occasionally, streaks of blood are
brought up with the hard coughing. In bad cases there
is cold sweat of the chest, or even of the whole body;
and in unfavorable cases extreme prostration, rattling,
suffocative breathing and a livid complexion. The
duration of the disease is from a few days to fifty or
sixty years.

CAUSES.—The disease comes on, at first, from exposure
to cold. The more attacks one has, the less exposure is
required to produce another.

TREATMENT.—For the acute febrile symptoms of the
first stage, give a dose of *Acon.* every hour. If the chill
is marked, *Gels.* is preferable;—dose, a drop every hour.

If the cough is dry, with much soreness and pain,
Bry. will be useful.

If the cough is dry and spasmodic, use *Phos.*

If the cough is hollow and ringing, use *Spongia.*

If the expectoration is profuse, thick and rather easily
detached, give *Tart. emet.* This remedy is especially
useful in the suffocative stage, to promote the expulsion
of the large quantities of mucus, which obstruct the
breathing.

When the mucus is tough and difficult to detach, *Kali bichrom.* is an excellent remedy.

The diet should, in acute cases, be very light. Mucilage of *Gum Arabic* or of *Slippery Elm Bark* makes an appropriate drink. Cold water or toast water may be allowed.

In chronic cases, especially with the aged, full, nutritious diet, and, in some cases, stimulants are required. The room should be warm. Hot poultices applied to the chest are very valuable adjuvants to the cure.

BRUISES.

The application of a lotion of *Arnica*, one part of tincture to six parts of water, generally prevents the black and blue appearance which might follow a contusion. Apply on soft linen, lint or cotton.

If the skin is broken, a cerate of *Calendula* is preferable. *Cosmoline* alone, forms an excellent dressing.

Arnica may be used internally, with great benefit, in cases of blows upon the head. Its effect is to contract the blood-vessels and prevent inflammation.

BUNIONS.

A bunion is a corn on a large scale, caused in a similar manner, having a similar structure, and requiring similar treatment. There is a thickened, inflamed condition of the *bursa*—water-sack—of the joint of the toe, which causes deformity of the foot.

CAUSE.—Narrow-soled, high-heeled boots.

TREATMENT.—A large, soft shoe should be worn and the foot allowed to rest, being kept as much as possible in an elevated position. The use of *Arnica*, locally, will greatly aid in dispersing the inflammation. A lotion made from the strong tincture, one part to five of water, or the cerate, may be used.

BURNS.

One of the best applications for burns is *Cosmoline*, spread thickly on cotton batting, and changed once in two to four days. If this is not at hand, a *raw potato*, scraped fine and bound loosely upon the surface, makes a very good dressing. The main points in the treatment are, to keep a uniform temperature of the part, and avoid the contact of all irritating substances.

CANCER.

A Cancer is a malignant tumor, that is, one which will probably return if it is removed, and finally wear out the vital energies of the patient, if some other disease does not come to his relief. Many benign tumors are called, by ignorant quacks, Cancers; and their abominable caustics and Cancer plasters, often gain a reputation for curing Cancers when they have only produced an ugly sore, which has after a long time healed spontaneously. If the existence of Cancer is suspected, go to a good surgeon and get his opinion and advice. Never trust a " Cancer Doctor."

CANKER.

DEFINITION.—By this is meant, small, superficial, whitish patches on the mucous membrane of the mouth or throat. It should be distinguished from the ulcerations which are found in *Sore-Throat*.

CAUSES.—These ulcerations are usually produced by an acid condition of the natural fluids of the mouth. This acid state is the result of a form of indigestion.

TREATMENT.—The local treatment is palliative and consists in the use of alkalies. The best is the Bi-Borate of Sodium, commonly known as *Borax*. Three grains of the first decimal trituration every hour will, in most cases, be enough to counteract the effects of the acid. If the ulceration is deep and indisposed to heal spontaneously, there may be a fungous parasite which can be destroyed by the application of a solution of *Carbolic Acid* of the strength of one to ten. This may be applied by dipping the end of a match into the liquid, and immediately placing it on the Canker. A wash of *Hydrastis*, one part tincture to ten parts water, is a valuable stimulant in such cases.

The constitutional treatment consists in improving the digestion. The food should be plain and mostly farinaceous; stimulants should be avoided. *Merc. cor.* may be administered, internally, three or four times a day.

CARBUNCLE.

A Carbuncle differs from a boil in its slow progress, in the greater constitutional disturbance which it produces, and in its having more than one opening.

TREATMENT.—The local treatment consists in applying, in the first stage, warm water dressing to relieve the pain, and freely opening the Carbuncle when matter is formed. Give internally *Arsen.* and *Sulphur.*

CATARRH.

DEFINITION.—This is a general name for an inflammatory condition of any mucous membrane; but in common language, is mostly limited to the affection of the mucous membrane of the air passages. The word is used in the latter sense in this article. For convenience, we will consider the disease first in the acute and afterwards in the chronic form.

ACUTE CATARRH.

SYMPTOMS.—Acute Catarrh includes "Cold in the Head," and "Cold on the Chest." The latter is treated of under the head of "Bronchitis."

A Cold in the Head generally begins with sneezing and watering of the nose and eyes, followed by the appearance of a thick, whitish discharge from the nose, which may be accompanied by considerable swelling and soreness of the nasal passages. The inflammation often travels down the air passages to the smaller bronchial tubes, when it is said, in common language, to be settled on the lungs. In bad cases the inflammation sometimes extends from the nostrils upward to the frontal sinus, causing headache; or it may extend from the throat, through the Eustachian tubes, to the middle ear, causing earache, noises in the ears and even deafness. Chilliness, followed by hot skin, quick pulse, headache and other symptoms of fever, marks the onset of the disease, especially in children.

CAUSES.—Exposure to cold when the body is insufficiently protected by clothing, and the blood is not in vigorous circulation. Colds are frequently taken immediately after exercise, when the strength of the circulation is diminished and the skin is covered with moisture. A cold is often taken by standing or sitting in a draft of cold air, when the circulation is inactive. The practice of allowing the clothing to dry on the body, when one is not in active exercise, is dangerous to health.

TREATMENT.—At the onset of a cold, if there is fever, take *Gels.*, in drop doses, every half hour for four or five hours, and keep the body well covered to promote perspiration. A warm foot-bath is a useful adjuvant to the sweating process. This treatment will, in many cases, abort an attack of acute Catarrh.

If the attack has gone too far or is too severe to be cured by this treatment, the following medicines will be found useful, as they are indicated by the symptoms.

Acon., if there is quick pulse, heat of skin and restlessness, a dose every hour, or every two hours in alternation with one of the following remedies :

Arsen., if there is a watery secretion from the nose, dry throat, thirst, chilliness, cold, sweat and general weakness.

Bryon., if there is soreness or sharp pains in the chest, with dry cough.

Kali. iod., if there is profuse flow of water from the eyes and nose with stuffed feeling in the chest and wheezing.

Tart, emet., if there is accumulation of thick mucus in the chest.

CHRONIC CATARRH.

This is generally the result of a succession of attacks of acute Catarrh. When the disease assumes this form, the discharge from the nostrils becomes thick, and contains pus as well as mucus. Sometimes the discharge falls back into the throat. In other cases it forms, in the nostrils, thick crusts which obstruct the breathing. In bad cases the discharge has an offensive odor. As in acute Catarrh, the discharge may extend to the frontal sinus, the ears or the bronchial tubes.

TREATMENT.—*Kali bichrom.* is exceedingly useful if there is hoarseness with tough, stringy mucus and coated tongue.

Merc. cor. is an efficient remedy if there is a profuse secretion of yellowish mucus, ulcerated Sore-Throat or Inflammation of the Eyes. The third decimal trituration, used as a snuff, has been found curative in some of the worst cases, where the disease was confined to the nasal passages. A small pinch is snuffed up, morning and night, for several weeks.

The use of the nasal douche is too dangerous to be generally recommended.

Arsen. and *Kali iod.* may be used, as in acute Catarrh, when the symptoms correspond to the indications given under that head. The most obstinate cases cannot be cured without a radical change of climate.

CHAFING.

This usually occurs in fat infants, sometimes in adults. The most frequent localities are, between the thighs, in the groin and on the neck.

TREATMENT.—Wash the parts often with warm water, using only the best toilet soap. Dry carefully, after washing, and apply a solution of *Glycerine and Tannin*, one drachm to the ounce. If this does not cure in a reasonable time, dust the parts, after every washing and drying, with *Lycopodium seeds*, or with the fine powder of *Oxide of Zinc*. *Cosmoline* is a sufficient remedy, in mild cases.

CHAPPED HANDS.

This affection is usually caused by exposing the hands to cold when they are wet. Persons subject to chapped hands, should be careful to have them covered when going into the cold.

TREATMENT.—Washing the hands in bran-water is often sufficient to restore the healthy condition of the skin. Bran-water is made by stirring a small handful of wheat bran into a basin of cold water. Bad cases may require the use of Calendula lotion of the strength of one to ten. "*Calendula Jelly*" made by adding Calendula to a hot solution of gelatin, is an excellent application.

CHICKEN POX.
(VARICELLA.)

This is a specific, contagious disease, peculiar to children.

SYMPTOMS.—For the first two or three days, there is fever, loss of appetite and general lassitude. On the second or third, a number of pimples appear on the body. These pimples soon become vesicles by filling with a watery fluid, and in two or three days point in the

center, burst and dry up. Successive crops of vesicles
appear for five or six days. The whole number of
vesicles varies from fifty to two hundred. The size
varies from that of a pin-head to that of a pea. The
larger vesicles, after shriveling, form a crust which lasts
four or five days. The disease is usually developed in
from thirteen to seventeen days after exposure.

TREATMENT.—Light diet, principally milk, and pro-
tection from cold.

CHILBLAINS.

These are the secondary effects of frost-bites. The
parts most frequently affected are the toes, the outside
of the feet, the heel, the ears, the nose and the fingers.
Parts, once partially frozen, often remain for weeks,
months or years, in a weak, congested state, ready to
swell and itch when excited by sudden changes of
temperature.

TREATMENT.—*Arnica* lotion or *Arnica* cerate, applied
locally, will mitigate the burning and swelling.

If the itching is very annoying, a lotion of *Cantharis*,
using one part of the strong tincture to ten of water,
is a soothing application.

The internal use of *Sulphur* will help to eradicate the
predisposition.

CHLOROSIS.

This is an affection peculiar to girls between the ages
of thirteen and twenty years.

SYMPTOMS.—The prominent symptoms of this disease
are: pale skin, puffiness of the eyelids, loss of appetite, or
perverted appetite, aversion to society, irregularity

(generally suppression) of the monthly discharge, and general loss of flesh, strength and vigor.

TREATMENT.—For the *anæmia* or thinness of the blood, give *Ferrum phosphoricum*, first decimal trituration, two or three grains after each meal.

For the nervous symptoms, *Ignatia*, *Cham.*, *Coffea* and *Acon.* are the chief remedies. See Materia Medica.

For the menstrual disorder, if there is irregularity, with frequently flushed face, and shifting pains in various parts, give *Puls.* If there is scanty, delayed or suppressed menstruation, give a dose of *Macrotin* every night. Out-door exercise and cheerful companionship are very important means in the prevention and cure of this complaint.

CHOLERA.

(SIMPLE CHOLERA.)

DEFINITION.—An acute catarrhal inflammation of the stomach and intestines.

SYMPTOMS.—Nausea, vomiting, purging of bilious or watery fluids, thirst, coldness, and sometimes cramps of the legs and abdomen.

TREATMENT.—If there is coldness and prostration or cold sweat, give two or three drops of the strong Homœopathic tincture of *Camphora*, strength one-fifth, on sugar, every hour.

If there is, in addition, vomiting and purging, give also *Verat. alb.*, once an hour.

If thirst predominates, give *Arsen.*

It is well to wrap the legs and abdomen in hot flannel. Vigorous rubbing with the hands may help to relieve the cramps of the legs.

CHOLERA ASIATICA.

(MALIGNANT CHOLERA.)

DEFINITION.—A form of Cholera produced by a specific poison, and characterized by the appearance of a fluid resembling rice-water, in the vomit and stools.

SYMPTOMS.—Sudden prostration of strength; coldness of the surface with great internal heat and thirst; cramps in the thighs, legs, toes and fingers; cold tongue and breath; vomiting and purging of fluid resembling rice-water. In the advanced stages, the pulse is scarcely perceptible; the eyes are sunken; the face is pinched; the voice is reduced to a hoarse whisper; there is extreme restlessness and thirst, with cold, clammy sweat.

TREATMENT.—*Camphora* is the main remedy in the treatment of this dreaded disease. Rubini's tincture, composed of equal parts of Camphor gum and pure, strong Alcohol, is the preferable preparation. This is two and one-half times the strength of the Homœopathic tincture, and four times the strength of the best tincture of the drug stores. The dose is one or two drops on a lump of sugar; to be frequently repeated.

Arsen. is a potent remedy in the cold stage of Cholera, when thirst, vomiting and blueness of the surface are prominent symptoms. *Verat.*, when diarrhœa predominates, and *Cuprum oxydatum*, sixth decimal trituration, when cramps of the extremities are troublesome.

CAUSE.—The germ theory of Cholera is now almost universally accepted by learned physicians. The Cholera-germ, although its form is not yet distinguished from those of other microscopic organisms, is believed to belong to the class known as *Protomycetes*. These are small, rounded, living bodies, which multiply with the most astonishing rapidity when placed in favorable conditions. The Cholera-germ can live in air or water, but its propagation takes place mostly in the fluids and secretions of the human body.   Drinking-water is the great channel through which the disease enters the human system.

PREVENTION.— *Carbolic Acid* is the best commonly known *anti-mycetic* agent.   The most convenient form for use is a solution of one part of the pure crystals in twenty parts of water, which is all the water will take up.   Where Cholera exists, this solution should be freely used, in chamber vessels, water closets, privies and sewers, and in every place where the secretions of Cholera patients may lodge.

To disinfect the air of the room, the fumes of burning *Sulphur* are the most thorough and convenient means known.   *Chloride of Lime* is far inferior to *Carbolic* and *Sulphurous Acids* as above used.   Drinking-water is disinfected by boiling, and clothing, by boiling water, by baking or by burning.

CHOREA.

(ST. VITUS' DANCE.)

DEFINITION.—An irregular convulsive action of the voluntary muscles, especially of the face and limbs.

SYMPTOMS.—This disease generally commences in the left arm, making it difficult to grasp any object with the left hand.   By degrees, the voluntary muscles of the whole body become affected, the limbs jerk about in every possible direction, and the face makes all sorts of grimaces and contortions, against the will and much to the annoyance of the patient. Children, between the ages of five and fifteen, are most subject to the affection; and girls are much more frequently attacked than boys. It occurs, occasionally, in pregnant women.

CAUSES.—A weak state of the system and a nervous organization predispose to the disease. It is frequently brought on by fright, as by telling a child that his head will be taken off. Cases of Chorea have been produced by "contagion of the eye," that is, by seeing the disease.

TREATMENT.—The disease generally disappears, spontaneously, in the course of a few weeks or months; hence it is that about a hundred different remedies have gained the reputation of curing St. Vitus' Dance.

The treatment is mainly moral and hygienic. The peculiar movements of the patient should be noticed as little as possible, and he should be encouraged to use the muscles, by giving him eggs and other fragile articles to carry. To strengthen the lower limbs the patient may be taught to walk on stilts. Plenty of good food, and rest in bed, are necessary to improve the general strength. Remove, if possible, all morbid conditions of the system which may tend to aggravate the disorder. If there is constipation, give *Nux.* If there is a pale, bloodless condition, give *Ferrum phosphoricum*, first decimal trituration, two grains after every meal. If there are symptoms of worms, give *Santonine.* If there is delayed menstruation, give *Puls.* or *Macrotin.*

COLD,—EFFECTS OF.

For cold in the head and cold on the chest, see "Catarrh."

For the muscular soreness and stiffness caused by exposure to cold, take, in the first stage, *Gels.*, in drop doses, every half-hour to every hour, and wrap the affected parts warmly to promote perspiration.

If this does not cure, and the pains become acute, take *Acon.* and *Bry.*

For lame back, take *Macrotin* three or four times a day.

See also "Neuralgia," "Inflammation of the Throat," etc.

COLD FEET.

Coldness of the feet is usually due to insufficient exercise. Walking, jumping the rope, and other exercises, are beneficial. It is well, also, to wash the feet every morning with, but not in, cold water, and rub them vigorously afterward.

If the hands and face are hot, give *Sulph.* every night and morning.

Delicate, nervous females, may receive more benefit from *Puls.* or *Sepia*, sixth decimal trituration.

If the feet are constantly sweaty, *Silicea* is the remedy. Two doses a day is sufficient.

COLIC.

This is an acute, cramping pain in the stomach or intestines, usually in the colon, the immediate cause of which is a violent contraction of the muscular fibres of some portion of the alimentary canal.

CAUSES.—The remote causes are indigestion, exposure to cold, worms and lead poisoning.

SYMPTOMS.—Ordinary Colic usually comes on without warning. The pain is just above the navel or thereabouts. The attack is often accompanied by nausea, vomiting and flatulence. The bowels are usually confined at the time of the attack. The pain is generally relieved after a free, spontaneous evacuation of the bowels. The affection is distinguished from inflammation of the bowels, by the fact that pressure relieves rather than increases the pain, as in the latter disease ; by the quiet state of the pulse and the coolness of the skin.

TREATMENT.—The patient should go to bed and be kept as quiet as possible. Warm flannel, bottles of hot water or bags of hot bran, may be applied to the abdomen, to relax the muscular spasms.

If there is constipation, give *Nux.*

If there is diarrhœa, give *Verat.*

If there is vomiting, give *Ipecac.*, in addition to one of the above ; or, if it can be readily obtained, *Dioscorea villosa*, given alone, in drop doses, will be found quickly curative.

Chamomilla is specially useful in the colic of infants.

The injection of a pint of tepid or quite warm water, will in many cases relieve the pain. See " Poisoning by Lead."

GASTRALGIA, Stomach Colic or Heartburn, is almost always the result of eating some indigestible food. It may be a strawberry, a teaspoonful of honey, an egg, or some other article of diet which the peculiarities of the

4

patient's digestive apparatus forbid him to eat. In mild cases, *Puls.* is a very efficient remedy. A practitioner of considerable experience, has called my attention to the nut of the sweet almond, as a cure for this complaint. The person subject to this affection, is to carry in his pocket a supply of the nuts, and on the approach of an attack eat one or a portion of one of the nuts. The relief is said to be almost magical. *Physostigma renenosum*, third decimal trituration, has proved in my practice a valuable remedy.

CONSTIPATION.

CAUSES.—Sedentary habits; the use of too fine or concentrated food; the poisonous action of lead; the habitual use of purgative medicines; irregular habits.

The following extract is from " The Stepping Stone," by E. H. Ruddock :

"A tendency to costiveness is not so grave a symptom as many persons suppose it to be ; indeed, individuals thus predisposed, generally live long, unless they injure themselves by purgatives ; while those who are subject to frequent attacks of diarrhœa are soon debilitated, and seldom attain old age. The common idea that aperients contribute to health, not only in sickness, but occasionally in health, and that impurities are thereby expelled from the body, is erroneous.

The fallacy of this notion may be easily demonstrated. Let purgatives be taken for a week, and however good may have been the state of health previously, at the termination of this period all sorts of impurities will be discharged. As this is an invariable result, even in the case of those who have never been ill, it proves that impurities are produced by these drugs."

Daily evacuation of the bowels should be the rule in youth and middle age; in infancy the movements may be two or three times a day, and in advanced age once in two or three days.

TREATMENT.—*Nux.* is indicated by the following symptoms : dull headache, dizziness, irregular action of the bowels, disturbed sleep, or the effects of over-eating, dissipation or sedentary habits.

Bryonia is preferable if there is slight fever, and soreness of the bowels.

Podoph. is useful when there is yellowish skin and eyeballs, with coated tongue and flatulence.

Sulphur should be given if there is chronic constipation, with piles, or burning, or itching of the anus.

Regular exercise, regular food, and a regular time for the movement of the bowels, are extremely important in the prevention and treatment of this disorder. Indian-meal mush, oat-meal mush, with molasses, Graham bread, Indian corn bread, ripe fruits and a variety of vegetables, should form a large portion of the diet. Cold water should be drunk freely, and tea, sparingly or not at all.

If there is a large accumulation of fæces in the rectum, it may be removed by injecting a pint of tepid water ; or, if the obstruction is obstinate, the same quantity of warm castile soap suds.

CONSUMPTION.

(TUBERCULAR CONSUMPTION OF THE LUNGS.)

DEFINITION.—The most common form of Consumption is the tubercular. Tubercles are small, rounded bodies, composed of morbid substances, which are deposited in the tissues of the lungs and various other organs of the body. They may remain dormant for several years after their deposit, and give scarcely any sign of their presence; but when the patient reaches the age of twenty years, or thereabouts, they generally take on active inflammation and soon disintegrate, soften, and are discharged, leaving ulcerated cavities which seldom heal.

SYMPTOMS.—The early symptoms of the inflammatory stage are : short, dry, hacking cough ; shooting pains through the upper part of the chest ; diminution of appetite, especially for fats ; shortness of breath ; evening fever, shown by flushing of the cheeks, thirst and quickening of the pulse and disturbed sleep.

Spitting or coughing up blood, is generally the first alarming symptom.   Hæmorrhage from the lungs comes mostly after exercise, loud talking or prolonged laughing; the blood is bright red and comes with a choking sensation, filling the mouth suddenly with pure blood. Hæmorrhage from the bronchial tubes comes after hard coughing, in strings or streaks, generally mingled with tough mucus.   Hæmorrhage from the stomach comes with retching or vomiting, and the blood is dark and thick.

In the course of the disease the expectoration becomes copious, thick and yellowish; the sputa sink in water unless they are of a frothy consistence ; the fever increases in violence ; the breathing becomes shorter, the pulse quicker ; night sweats and occasional diarrhœa supervene ; the patient rapidly loses flesh, and in the great majority of cases, death ends the scene in two or three years from the appearance of the first symptoms.

CAUSES.—In most cases, the disease is inherited ; in many it is acquired by infection ; and in a few, perhaps, it is generated *de novo*.

PREVENTION.—The marrying of persons known to be infected with the disease, should be strictly prohibited. Consumptives should be isolated as much as possible, to prevent the poisoning of the healthy by the entrance of tuberculous matter with food, drink or air. Persons in whom the disease is hereditary should live as much as possible out of doors. If the weather is too severe, they should be removed to a mild, dry and equable climate. The most nutritious food should be used, in order to keep up the general strength, and all exposure to sudden cold, and all excesses should be avoided. Those who pass the age of twenty-five without a development of the disease may begin to cherish a hope of escape, but extreme care and vigilance should be exercised till the patient has passed the age of thirty-five.

TREATMENT.—Hæmorrhage from the lungs may be checked, in some degree, by the use of *Acon.* *Hamamelis,* and *Erigeron Canadense,* one or the other, in five drop doses of the mother tincture, are the best remedies to stop the flow of blood. The dose may be repeated every five minutes. The distressing cough of the first stage may be relieved by the administration of *Phos.* The second decimal trituration of *Sanguinarin,* made with one-third extract of *Licorice,* is an excellent powder to allay the irritation which produces the cough. *Tart. emet.* is the best cough remedy in the later stages.

Acon. and *Bryon.* should be used to allay the fever and the shooting pleuritic pains.

Arsen. is useful in case there is a suffocated feeling in the chest, and will also help in the profuse night sweats and the wasting diarrhœa.

The diet should be of the most nutritious character. Cod Liver Oil, the richest portion of milk, the yolk of eggs, and pounded beef, are among the best articles of diet, but no one food should be used so long as to produce a distaste for it. Stimulants may be used sparingly but only when there is no fever, and always with the food.

CONVULSIONS.

DEFINITION.—By this term is here meant Infantile Convulsions or Fits of Infants.

SYMPTOMS.—Twitchings of the muscles of the face; rolling up of the eyes; irregular, sighing breathing; clenched hands with inturned thumbs; throwing back the head. The fit usually lasts one or two minutes, and may be succeeded by others at longer or shorter intervals.

CAUSES.—Irritation of the brain from teething, or from the presence of worms in the alimentary canal; from Dropsy of the Brain, or from the fever which accompanies nearly all the acute diseases of infants, fright. suppressed eruptions and indigestion.

TREATMENT.—If there are signs of congestion or inflammation of the brain, such as flushed face and fullness of the fontanelles, give *Bell.*

Cham. is the remedy if there are only slight twitchings and starting in sleep, or if one cheek is red and the other pale.

If there are signs of worms, give *Sant.*

On the appearance of Convulsions, the child should be stripped and put into a vessel of warm water, of the temperature of about 100° F. This will generally relax the rigidity of the muscles. The child should not be left in the bath more than five minutes. A cloth wrung out of cold water may be applied to the lower and back portion of the head and frequently changed.

CORNS.

A Corn is a hard, thickened, inflamed condition of the skin of some portion of the foot. A hard Corn has generally a whitish, horny center, and sometimes a kind of root which sends out branches. When a Corn comes between the toes it is kept in a moist condition, and is softer in texture.

CAUSES.—Narrow-toed, high-heeled boots or badly fitting socks.

TREATMENT.—Remove all pressure from the Corn by procuring and wearing only boots or shoes or stockings which fit the feet, giving plenty of room for the toes to spread. Pare down the Corn with a sharp knife till no pain is felt on pressing it with the finger, and apply a felt corn-plaster. *Calendula cerate* may be applied to soften the Corn, and to prevent the inflammatory action which causes it to return.

COUGH.

Cough is usually only a symptom of some graver trouble, such as Consumption, Bronchitis or Dyspepsia. Sometimes, however, it seems to be merely a habit.

TREATMENT.—Spasmodic cough with dry or inflamed throat, requires *Bell.*

Dry, hard, painful cough with stitches in the chest, requires *Bry.*

Short, hacking cough with tight feeling in the chest, and frothy, rust-colored sputa, requires *Phos.*

Loud, hollow, ringing cough, requires *Spong.*

Nux. is useful if there is fullness of the stomach, and *Pod.* if there is yellow skin with coated tongue and constipation.

Puls. will often cure a catarrhal cough when there is running of the eyes and earache.

CRACKING OF THE LIPS.

TREATMENT.—Apply, daily, *Calendula Jelly.*
See "Chapped Hands."

CRAMP.

DEFINITION.—A violent, involuntary contraction of a few of the voluntary muscles.

CAUSES.—Cramps of the muscles of the stomach and bowels are caused by the presence of irritating bodies, as poisons, indigestible food or worms. Cramps of the legs and arms occur in Cholera. They may be produced by exposure to cold, as in bathing, or they may be the result of a deficient supply of blood to the parts.

TREATMENT. — Consult the articles on "Cholera," "Poisons" and "Worms." Cramp in the extremities, occurring in persons of good health, may be relieved by rubbing, spatting or warming the affected part. The general preventive treatment consists in the use of the cold bath, out-door exercise and other hygienic means.

CROUP,—MEMBRANOUS.

DEFINITION.—An inflammation of the larynx in which a tough, stringy, yellowish-white exudation takes place upon the mucous membrane.

CAUSES.—Exposure to sudden changes of temperature, especially to north and north-east winds. The disease is not contagious, although it very closely resembles Diphtheria.

SYMPTOMS.—The symptoms of the premonitory stage are : hoarseness and loss of voice, with a hollow, ringing cough, which soon loses its tone. The hoarse, barking sound often changes suddenly from a deep base to a high falsetto. The respirations are thirty to the minute. The child is soon obliged to sit up in order to breathe. The inspirations are accompanied by a whistling or sawing sound, and the expiration has more of a snoring sound. Suffocative attacks come on at intervals, in which the little patient shows the greatest anxiety, the eyes protrude and the hands clutch at the throat as if to remove some obstruction. These choking-spells almost always become more frequent and more severe until one of them ends a most heart-rending struggle for life, or until the patient dies of exhaustion.

Fortunately, true Croup is of much rarer occurrence than Spasmodic Croup; and, fortunately again, one attack of the disease gives immunity from subsequent attacks. Prof. Steiner, of the Children's Hospital of Prague, who has seen one hundred thousand cases of Membranous Croup, says he has never yet met a single recurrence of the disease. Spasmodic Croup, as a rule, often repeats itself.

TREATMENT.—*Acon.* and *Kali. bichrom.* are the most generally useful remedies.

Cold, wet cloths should be applied to the throat and frequently renewed. The inhalation of the vapors which arise from slaking lime, has been found to be the only reliable local treatment.

CROUP,—SPASMODIC.

DEFINITION.—A catarrhal inflammation of the larynx, producing paroxysmal attacks of difficult respiration.

CAUSES.—Exposure to cold and dampness. Infants and young children are most subject to the disease.

SYMPTOMS.—During the first day or two, the patient has slight hoarseness and a cough similar to that which accompanies an ordinary cold. On the second or third night, the child awakes from sleep with a sense of suffocation; loud, harsh inspiration and a dull, barking, dry cough, alternating with a doleful cry. The voice is harsh and toneless, and there is great restlessness and anxiety. The temperature of the body is but slightly raised.

With the proper treatment these alarming symptoms subside in one or two hours, and the patient goes into a quiet sleep, which may last till the following morning. Through the day he is comparatively well, but the attack may occur the second or even the third time.

TREATMENT.—Give *Gels.* in drop doses every half hour during the paroxysm. During the remission, give *Acon.* and *Spongia*, alternately, one hour apart; or if the secretion of mucus is profuse, give *Acon.* and *Tart. emet.* in the same way. A cloth wrung out of hot water may be applied to the throat and upper part of the chest. This disease, though alarming, is seldom fatal.

CUTS.

A Cut should be closed up as soon as possible and the edges held together by stitches of silk or linen, or by adhesive plaster.

Excessive hæmorrhage may be checked by holding the cut portion as high as possible and pressing firmly upon the edges. If a spurting stream is observed which cannot be stopped by these means, press more firmly and hold the part for some minutes; then stitch or plaster the wound.

The less blood there is in a cut the sooner it will heal.

If the skin is torn or bruised, the application of *Calendula Cerate* will be found agreeable, and will facilitate the healing process.

DANDRUFF.

DEFINITION.—A superficial inflammation of the skin, characterized by the excessive growth and falling off of whitish, bran-like scales.

SYMPTOMS.—The disease occurs mostly upon the hairy scalp, but often at the roots of the eyelids and on other parts of the body. There is an unnatural dryness and heat of the skin, and at times considerable itching.

TREATMENT.—*Arsen.* is the specific remedy. The scalp should be washed frequently, with pure, soft water. To facilitate the removal of the scales a little *Borax* may be added to the water. After each washing it is well to apply a small quantity of some oily substance to prevent the itching. *Cosmoline* is an excellent dressing for this purpose.

DEBILITY.

Debility is a symptom of some grave, constitutional disorder. It occurs in Pulmonary Consumption, Dyspepsia, Diabetes, Leucorrhœa and Spermatorrhœa. See the treatment under these heads.

DIABETES.

(DIABETES MELLITUS.)

DEFINITION.—A constitutional disease characterized by an excessive discharge of pale, sweet and heavy urine, containing grape-sugar.

SYMPTOMS.—The principal symptoms are: debility; loss of flesh; intense thirst; sinking feeling at the stomach; constipation; red, dry tongue; dry, harsh skin; a lowering of the body temperature from two to five degrees Fahrenheit; the passage of a greatly increased quantity of pale, sweet, heavy urine. The specific gravity of the urine of Diabetes varies from 1.025 to 1.050, that of water being 1.000, and that of healthy urine, from 1.018 to 1.030. The test for sugar may be made by a practical chemist.

CAUSES.—Little is known of the remote cause of this almost incurable disease. There is a defect in the chemical process by which the sugar and starch of the food are appropriated to the nutrition of the body. The normal process is interrupted, or switched off, at the point where grape-sugar is produced, and the excess of this substance in the blood is carried off by the kidneys.

TREATMENT.—The patient must abstain, as much as possible, from eating substances which contain grape-sugar or which may be converted into grape-sugar. Such substances are *ordinary bread* and all preparations from *ordinary flour, fruits, potatoes,* and *vegetables,* except a little *cabbage* and *lettuce.* The bulk of the food should be *animal food,* including *fats, milk* and *eggs,* and *bran-cakes.* The bran should be washed, to free it from starch, and finely ground, to prevent its producing indigestion. I cannot do better than quote the formula of Dr. Camplin, given in his Monograph on Diabetes :

" Take a quantity of wheat bran, (say a quart), boil it in two successive waters for a quarter of an hour, each time straining it through a sieve ; wash it well with cold water on the sieve, until the water runs off perfectly clear. Squeeze the washed bran in a cloth, as dry as possible, then spread it thinly on a dish, and place it in a slow oven. When it is perfectly dry and crisp, it is fit for grinding into fine powder.

The bran thus prepared is ground in the mill for the purpose, and must be sifted through a wire sieve of such fineness as to require the use of a brush to pass it through, and what remains on the sieve must be re-ground, till it is sufficiently soft and fine.

To prepare a cake, take of this bran powder three or four ounces, three new-laid eggs, one and a half or two ounces of butter, and about half a pint of milk. Mix the eggs with a little of the milk and warm the butter with the remainder of the milk ; stir the whole well together, adding a little nutmeg and ginger, or any other spice that may be agreeable. Bake in small tins (pattipans which must be well buttered), in a rather quick oven, for about half an hour. The cakes when baked should be a little thicker than a captain's biscuit.

These cakes may be eaten with meat or cheese for breakfast, dinner and supper, and require a free allowance of butter ; and the cakes are more pleasant if placed in the oven a few minutes before being placed on the table.

When economy is an object, when a change is required, or if the stomach cannot bear butter, the cakes may be prepared as follows : Take of the prepared bran four ounces, three eggs, about twelve ounces of milk, with a little spice and salt, to be mixed and put into a basin, (previously well buttered). Bake it for about an hour ; the loaf may then be cut into convenient slices and toasted when wanted ; or, after slicing, it may be re-baked, and kept in the form of rusks.

Nothing has yet been discovered of equal utility to these bran cakes, combining as they do, moderate cost with freedom from starch, and sufficient pleasantness as an article of food."

Uranium nitricum, second decimal trituration, three grains before each meal and at bed-time, has been found a remedy of great power in the early stages of this disease. *Phosphori Acidum*, first decimal dilution, two drops three times a day, has been strongly recommended. *Arsen.*, *Nux*, etc., may be required as intercurrent remedies.

DIARRHŒA.

DEFINITION.—Frequent fluid evacuations from the bowels, not accompanied by straining in the rectum and not containing blood.

SYMPTOMS.—The constitutional disturbance may be so slight as not to be noticed, or it may give rise to all the symptoms mentioned under "Cholera simplex."

TREATMENT.—If the cause is *indigestible food*, *Puls.* and *Nux* are the usual remedies. If there is nausea, *Ipec.* or *Tart. emet.* are better.

If the cause is *exposure to cold*, *Acon.* should be given. *Dulcamara* is said to be a remedy for the effects of dampness or the disposition to take cold.

For Diarrhœa caused by *hot weather*, give *Verat.* if there is coldness and colicky pain, or *Arsen.* if there is prostration of the strength and thirst. *Iris versicolor*, second or third decimal dilution, or *Irisin*, second dec. trituration, will be found efficacious where there are nausea and headache.

For Diarrhœa caused by the irritation of *teething*, *Cham.* is the specific remedy.

For Diarrhœa from *wasting diseases*, give *Phos.*, or if there is also excessive sweating, *Camph*. *Nitri Acidum*, first decimal dilution in water, one drop every two hours, is a very useful remedy in such cases.

Chronic Diarrhœa may require *Sulph*. or *Ferrum phosphoricum*, first dec. trit., two grains after each meal.

The food should be light, and, in case there is pain, should be mostly limited to boiled corn starch or boiled rice, with a little milk and sugar. The application of heat to the bowels by means of hot flannel cloths or vessels of hot water, will assist to relieve the pain. Rest in bed, is in some cases necessary, and in all cases preferable.

DIPHTHERIA.

DEFINITION.—An acute inflammation of the mucous membrane of the throat, attended by the exudation of a membranous substance, which has the power of reproducing itself when placed upon a denuded surface.

CAUSES.—Diphtheria may originate in the body spontaneously, or it may be induced by contagion from other cases of the disease. The most accurate observers state that there are always present in the exudation, minute, living organisms, called *Round Bacteria* or *Micrococci*; and that if a portion of the exudation, con_ taining these organisms, be planted upon a raw surface they multiply with great rapidity. The germs of the disease may be carried in food, water, air or clothing.

SYMPTOMS.—There is in the beginning, slight fever; loss of appetite and strength; some soreness and pain in the throat; swelling of the glands near the throat. In the first stage there is only a reddening of the surface of the affected parts, but within twenty-four hours may be seen small, yellowish-white, *elevated* patches, which, in mild cases, loosen and are thrown off in four or five days, but in severe cases increase in extent and thickness, and assume a grayish color. In these severe cases the fever rises rapidly and the patient becomes restless, and as the false membranes spread downwards the patient experiences difficulty of breathing. Strips of the false membrane may be thrown off by coughing, but it is formed again wherever the surface has been made to bleed and gradually extends downward to the bronchial tubes and chokes the patient to death.

The severer forms of the disease are more fatal in children than in adults. The lighter forms can sometimes scarcely be distinguished from ordinary ulcerated Sore-Throat. It should always be remembered that in Diphtheria the exudation lies on the *surface* of the mucous membrane, and, when removed, leaves it almost intact; while the white patches of ulceration may have raised edges, but when removed, they leave an *excavated* spot.

TREATMENT.—In the early stage, *Bell.* and *Merc. iod. rub.* are the most useful remedies known. These should be given, in alternation, an hour apart. In the milder cases the disease will thus be hastened to a favorable termination.

When the false membrane becomes grayish and thick, and the patient begins to show signs of blood-poisoning, *Chininum arsenicosum,* second decimal dilution, one drop every two hours, will prove of great service in keeping up the patient's strength until the poison can be eliminated.

When the false membrane shows a tendency to travel rapidly downward, *Kali bichr.* should be given, and inhalation of the steam from slaking lime may prove useful, as in Membranous Croup.

Bromium and *Sulphuris Acidum* may be required in bad cases.

The diet should be of the most nutritious kind. Milk, beef-tea, eggs beaten up in milk, or iced milk, may be given as freely as the patient can swallow them. The patient may be allowed to hold pieces of ice in the mouth till they melt.

The best gargle, in bad cases, is a solution of *Kali permanganicum*, two grains of the crude salt to one ounce of water.

DIZZINESS.

Dizziness may be a symptom of one of several different affections.

When it is caused by *indigestion*, it is generally accompanied by some sensation referable to the stomach; as nausea, eructations, fulness or "goneness" of the stomach, coated tongue or bad breath. In such cases give *Nux*, if there is constipation; *Bryon.*, if there are bitter eructations ; and *Ipecac.*, if there is nausea.

When it is caused by *nervous* exhaustion, give *Ignatia*.

When it is caused by swinging, or by riding on rough water, or traveling by rail, give *Nux*. *Cocculus Indicus*, third dec. dilution on pellets, is also a useful remedy in Sea-sickness.

When disease of the heart is suspected, dizziness should warn the patient that it is time to sit or lie down. *Arsen.* and *Cactus grandiflorus*, second dilution, in alternation, will help to relieve the dizziness of Heart-disease.

DROPSY.

DEFINITION.—This is a general name for an unnatural accumulation of water in the cavities or tissues of the body.

CAUSES.—Dropsy may be caused by chronic disease of the heart, chronic disease of the liver, acute or chronic disease of the kidneys, or, as in Dropsy of the Brain or of the Chest, by inflammation of the serous membranes.

SYMPTOMS.—Dropsy from disease of the *heart* begins in the feet and works upwards. It is generally accompanied by cough and difficulty of breathing.

Dropsy from disease of the *kidneys* may early be noticed under the eyes, and it begins about the same time in different parts of the body. It is generally accompanied by pain in the region of the kidneys, and scantiness of the urine. Whether the primary cause be in the heart or kidneys, the swollen parts " pit " when they are pressed upon by the finger.

If the cause is in the liver, the swelling usually begins in the cavity of the abdomen, and afterwards, commencing in the feet, works upwards in the same manner as in cases of heart or kidney disease.

Dropsy of the Brain,—Hydrocephalus—much oftener affects infants than adults. The early symptoms are, sudden crying out in sleep, heat of the head, grating of the teeth, fulness of the fontanelles. Later, the head enlarges, greatly out of proportion to the growth of the body ; the patient loses flesh, the intellect fails and convulsions, coma and death supervene.

Dropsy of the Chest,—Hydrothorax—is generally the result of chronic *Pleurisy.* The affected side of the chest swells, the spaces between the ribs puff out, and the breathing is short and quick.

TREATMENT.—*Arson*, is one of the best remedies for Dropsy of the *tissues* from whatever cause. The swollen feet and legs should be frequently rubbed with the hands, and kept in an elevated position. The wearing of elastic stockings is a source of great comfort, in such cases. *Apis mel.* is a valuable medicine in dropsy from *acute* disease of the kidneys.

The absorbtion of water from the cavities of the brain and *pleura* may be promoted by the administration of *Merc. iod. rub.*

DYSENTERY.

DEFINITION.—Dysentery is an inflammatory disease affecting the mucous membrane lining the lower portion of the bowels.

CAUSES.—Sudden exposure to cold, and improper food. The disease prevails mostly in autumn.

SYMPTOMS.—The attack usually begins with a chill, followed by a slight fever. The patient has a frequent desire to go to stool, but the passages are scanty and contain little more than slime and blood. A peculiar symptom of this disease, is the straining and pain in the lower portion of the bowels which *follows* the evacuations.

TREATMENT.—The patient should be kept perfectly quiet and in the reclining posture. The room should

be warm.   The food should be of a liquid or pulpy consistence.   Milk, gruel, boiled rice and boiled corn starch
may be given, frequently, in small quantities.   The griping pains in the rectum may be relieved by the injection
of boiled starch and warm water.

Merc. cor. is the specific remedy.   A good rule for
frequency is to give a dose after each movement of the
bowels.   If there is chilliness or fever *Acon.* should be
given occasionally.   If the discharges are of clear blood,
give *Erigeron Canadense,* one to five drops of the strongest
tincture every hour.   If the duration of the disease is
prolonged, give *Hydrastis Canadensis,* strongest tincture,
a drop every two hours.

DYSPEPSIA.

DEFINITION.—Imperfect or painful digestion, usually
in a chronic form.

SYMPTOMS.—The symptoms are of the most various
kinds.   Among them are, diminished or irregular appetite, bad taste in the mouth, eructation of sour or
bitter fluids, pain or fulness in the region of the stomach,
and palpitation of the heart.

CAUSES.—Imperfect mastication of food, irregular times
for meals, over-eating, eating of imperfect food, etc.

TREATMENT.—All known causes should be carefully
avoided.   The following are the principal remedies with
the indications for their use :

Nux—Fulness and tenderness of the stomach, after
meals ; sour eructations ; flatulence ; *dull,* or confused
feeling in the head ; attacks brought on by the use of
stimulants, or by sedentary habits.

Puls.—Acute indigestion from eating fats or pastry ; shifting pains.

Bryon.—Bitter taste ; bitter eructations ; stitching pain or soreness, on taking a deep breath ; constipation ; headache, worse on movement ; irritable disposition.

Carbo veg.—Bad breath ; heartburn ; rising of water from the stomach ; belching of wind.

Arsen.—Sour stomach ; pale, bloated appearance ; ulcerated mouth or throat ; thirst ; occasional diarrhœa.

Physostigma venenosum, third dec. trit.—Feeling as though there were a lump in the stomach, heat and fullness of the head.

DIET.—Every case of Dyspepsia has its peculiarities, and no general dietary can be prescribed. Anything which has been found by experience to disagree with the patient, should be used afterwards with caution. Fresh, tender animal food, cooked so as to retain the natural juices, should generally form a large share of the diet. Smoked, dried, or twice cooked meat, sausages, salmon, lobsters, cucumbers, raw vegetables, new bread and coffee, must generally be avoided. Boiled rice and fine food should be used when there is a tendency to vomiting. Graham bread and the different kinds of mush may be used, if the bowels are constipated.

EAR,—FOREIGN BODIES IN THE.

Children frequently get beads, beans, peas, pebbles or other substances in the external ear; and adults as well as children, occasionally get a live bug in the ear.

TREATMENT.—Small bodies may be removed by holding the head with the affected ear downward, and throwing a gentle stream of warm water into the ear, with a syringe. The looped end of a hair-pin is a very convenient instrument for the removal of larger bodies. The greatest care must be taken to make no inward pressure, and not to introduce the instrument more than five-eighths of an inch beyond the external surface. Bugs may be killed, before removal, by placing in the ear a piece of cotton saturated with *Alcohol* or *Chloroform*.

EPILEPSY.

DEFINITION.—A functional disease of the brain, characterized by periodical attacks of partial or complete loss of consciousness and muscular power. The severity of the attack varies from the most violent fits of convulsions to a slight dizziness or confusion which only the patient can recognize. The "convulsions of infants" are epileptic. See "Convulsions."

SYMPTOMS.—The warning symptoms are not always present. These are the same for each individual in every attack, but different patients have symptoms so various that no attempt will be made to enumerate them.

The symptoms of the attack are in the worst cases as follows : The patient falls suddenly, the eyes are found half open, the eyelids quiver, the face is red and swollen, the mouth foams, the jaws are closed convulsively with irresistible force, the arms and legs are thrown about, the hands are clutched and the body is bent backward. The duration of the attack in children is only a few minutes, in adults it may last two hours. After the attack the patient falls into a stupid sleep which lasts three or four times as long as the attack. The intervals between the attacks vary from a few minutes to several weeks. ,

CAUSES.—Teething is a very common cause in infants ; worms in the alimentary canal are less frequent causes. Fright sometimes produces an attack in children, and errors in diet or the use of alcohol in older persons.

TREATMENT.—During the attack, the patient should be laid upon a bed or on the covered floor, and restrained only enough to prevent his doing injury to himself.

After the attack, the patient should be allowed to sleep quietly as long as he will.

To prevent the recurrence of the attacks, in chronic cases, *Kali bromatum*, crude, in doses of five to fifteen grains, three times a day, has been found useful. This is the dose for an *adult*. The medicine may be taken dissolved in milk to cover the taste.

Bell. should be given, in the premonitory stage, if there is congestion of the face or headache. *Nux* is useful during the intervals between the attacks, to regulate the digestive functions.

ERYSIPELAS.

DEFINITION.—This is an acute inflammation of the skin and sometimes of the deeper tissues, accompanied by fever.

CAUSES.—Exposure to cold, hereditary predisposition, wounds and contagion.

SYMPTOMS.—In severe cases, the fever precedes the eruption for two or three days, but does not subside on its appearance. The fever is accompanied by frequent chills, thirst, nausea, coated tongue, sore throat and swelling of the glands of the neck.

The eruption is, at first, of a bright red color, but afterwards assumes a livid hue. There is a constant burning and itching sensation in the skin. In the course of the disease, the areolar tissue under the affected portion of the skin becomes inflamed, and sometimes pus is formed and discharged.

The face is most frequently affected, but the inflammation may attack any portion of the cutaneous surface. In some cases the swelling is so great as to obliterate the features of the countenance.

The duration of the disease varies from two or three days to as many weeks, depending much upon the treatment.

Erysipelas is always a serious disease, but is especially fatal to the aged and the intemperate. The principal danger lies in the metastasis of the inflammation from the face or scalp to the membranes of the brain.

TREATMENT.—The patient should be kept in bed. The

room should be well ventilated and cool, but not very light. Cooling drinks, especially water and lemonade, may be given frequently.

Acon. and *Bell.* in alternation, are the best remedies in the early stage, and *Bell.* may be of service in any stage.

Rhus. is a specific for Erysipelas, and may be given in alternation with *Bell.*, when the fever is not so high as to require *Acon.*

Cantharis lotion (made by mixing one part of the strongest tincture with ten or fifteen parts of water) is one of the best local applications. Moisten a piece of linen cloth in the lotion, and lay it, one or two folds thick, on the inflamed surface. The use of *astringent* lotions, is dangerous, if the disease is located near the brain. In most cases no external remedies need be applied.

EYE,—FOREIGN BODIES IN THE.

Particles of dust, small hairs, cinders and other substances which lie loosely under the lids, may easily be removed with a soft handkerchief, so folded as to make a sharp, rather stiff corner, moistened and used as a brush. The sac of the lower lid, may be exposed to view by simply pulling out the lid. The sac of the upper lid is made visible by pinching the lid between the thumb and finger and pulling outward. The lid must be pinched from side to side, not up and down.

To expose the sac more fully, it is sometimes necessary to turn out the upper lid. This may be done by laying a match or penknife blade across the lid about a quarter of an inch above the edge, and pulling the edge of the lid by the lashes, outward, upward and backward.

When foreign bodies have become so firmly imbedded in the mucous membrane, that they cannot be wiped off, they may be scraped off with the edge of a very sharp penknife, or dug out with the point of a needle. See " Inflammation."

FAINTING.

(See "Syncope.")

FELON.

DEFINITION.—A Felon is an inflammation of the fibrous sheaths of the bones of the fingers or thumbs. This is to be distinguished from the two other forms of whitlow, namely, the cutaneous, which is situated in the skin about the nail, and the subcutaueous, which is situated in the soft tissues under the skin.

CAUSES.—Impoverished state of the blood, and dipping the hands into hot water.

SYMPTOMS.—Deep-seated, aching, throbbing pain in the finger or thumb, followed by soreness, heat, swelling and the formation of pus, which burrows around the bone till it finds an easy vent, through the surrounding tissues. When the disease is not properly treated, the burrowing pus is liable to peel off the sheath of the bone, portions of which die and fall off.

TREATMENT.—Keep the skin and subcutaneous tissues soft by dipping the Felon occasionally in lye, or by dressing it with a poultice of soap. When the part begins to swell it should be lanced *deep and long*, laying bare the bone. *Silicea*, third decimal trituration, is the best internal remedy.

FEVER.

DEFINITION.—The main characteristic of Fever is an elevation of the temperature of the blood, from one to fourteen degrees above the normal standard of 99°, Fahrenheit.

SYMPTOMS.—These are, languor, weakness, chilliness, loss of appetite, quick pulse, and general heat.

CAUSES.—Either local inflammation or blood poisoning.

CLASSIFICATION.—Fevers have been variously classified, but the following arrangement seems the most scientific, as well as the most practical. Symptomatic fevers are those which accompany local inflammations. Idiopathic fevers are those which proceed from general blood-poisoning. By contagious fevers are here meant those which are communicated, directly or indirectly, from individual to individual. Non-contagious idiopathic fevers originate outside the body, generally in the soil. Semi-contagious fevers generally originate from external sources, but may, in rare instances, be communicated from individual to individual.

CLASSIFICATION OF FEVERS.

IDIOPATHIC,					SYMPTOMATIC,
CONTAGIOUS,		SEMI-CONTAGIOUS,	NON-CONTAGIOUS,		
Eruptive,	Non-Eruptive,				
Small-Pox,	Puerperal Fever,	Typhoid Fever,	Intermittent Fever,		Brain Fever,
Cow-Pox,	Mumps,	Relapsing Fever,	Remittent Fever,		Quinsy,
Chicken-Pox,	Whooping-Cough.	Yellow Fever,	Meningitis,		Pneumonia,
Swine-Pox,		Cholera,	Rheumatic Fever,		Pleurisy,
Measles,		Dysentery.	Gout.		Croup, etc.
Scarlet Fever,					
Typhus Fever,					
Erysipelas,					
Syphilis.					

ERUPTIVE FEVERS.

The diagnosis of the eruptive fevers may be facilitated by consulting the following table and the remarks underneath :

Name of Disease.	Period of Incubation.	Period of Fever.	Period of Eruption.
Scarlet Fever..................	7 days.	1½ days.	4 days.
Measles.......................	10 "	4 "	5 "
Small-Pox....	12 "	4 "	6 "
Chicken-Pox..................	15 "	2 "	4 "

The eruption of Scarlet Fever is preceded by vomiting and *sore-throat*. It appears first on the neck. The rash is fine and uniform, and turns pale when touched by the finger.

The eruption of *Measles* is preceded by *sneezing* and *watering of the eyes*. It appears first on the *face,* and begins to fade after four days. The eruption is in spots which disappear on pressure.

The eruption of *Small-Pox* is preceded by *intense head-ache, pain in the back,* and a great tendency to *vomiting.* It appears first on the *face,* in the form of small round pimples, which gradually enlarge and fill for four days, then pit in the center and become covered with crusts, which fall off after ten days. The number of the pimples, or pustules, varies from twenty to some thousands.

The eruption of *Chicken-Pox* is preceded by very slight febrile symptoms, and appears first on the breast and back. The pimples fill in three days, then crust over for five days, but seldom pit in the centre like those of Small-Pox.

INTERMITTENT FEVER.

DEFINITION.—This is a form of fever in which the febrile symptoms occur in paroxysms of considerable regularity, and between which are intervals of freedom from these symptoms.

CAUSE.—Residence in a locality where there is a drying up of damp soil. Marshes which are covered with low water at one season of the year, and which become dried by the sun's heat during another season, are the great breeding places of the miasmatic poison which produces Intermittent Fever. The turning up of the subsoil in newly cultivated countries, or the digging up of the subsoil to any great extent, even in old countries, often produces the miasm.

SYMPTOMS.—The paroxysm of Fever consists of three stages :

The *first* is the *chilly* stage. This begins with languor, yawning, stretching, and pains in the head, back and limbs. Then follows a sensation of coldness, beginning in the back and spreading to the surface. This sensation is, at first, occasionally relieved by intervals of warmth, but afterwards becomes constant. The actual tempera-ture of the blood, during the chill, is above the normal standard, but the surface of the extremities is cold. The duration of the chill may be from a few minutes to six hours, but the ordinary period is from one to two hours.

In the *second* or *hot* stage, a feeling of heat radiates from the center to the surface, and the body becomes hot and the extremities warm. The pains in the back and extremities are relieved, but the headache is increased. The face from being blue and pinched, becomes red and full. The patient complains of roaring in the ears and sparks before the eyes. There is thirst and restlessness, or even delirium. This stage may last from one to twelve hours, but generally continues from two to three hours.

The *third*, or *sweating* stage, comes on gradually. The headache is relieved, the pains in the limbs and other febrile symptoms disappear, and the body is covered with a free perspiration, which dries up so gradually that it is impossible to define the exact limits of this stage.

The duration of the whole paroxysm is usually from four to twelve hours.

In the *apyrexia* or intermission, there is an apparent return to health, with the exception of some dyspeptic symptoms, such as coated tongue, bad taste in the mouth, and want of appetite and strength.

The paroxysm is usually repeated every forty-eight hours, till the fever is broken, but in some cases it comes once in twenty-four hours, and in others once in seventy-two hours.

In some cases a *neuralgic headache* takes the place of the *chill* of the first stage, and in that form of malarial poisoning, called "Dumb Ague," a *periodical headache* is the most prominent symptom.

The secondary effects of the poison may continue for months or years after the chills and fever have ceased. Among these effects are : enlargement of the liver, enlargement of the spleen, and a condition of general debility, with paleness and loss of flesh.

PREVENTION.—If it is necessary to reside for a short time in a malarious district, go in winter rather than in summer ; select for residence the highest and driest parts of the country, and sleep in upper rooms facing the south, and have the windows closed at night. Avoid fatigue and all excesses in diet.

TREATMENT.—The palliative treatment consists in modifying the severity and shortening the duration of the paroxysm. *Gels.* is the most efficient remedy for this purpose. The dose for an adult is three drops every hour, during the chilly stage, and one drop every hour, during the hot stage.

The curative treatment consists in preventing the recurrence of the paroxysm and removing the effects of the poison. When the disease is fully developed, anti-periodic medicines are required to break it up. This class of remedies should be used only during the inter-mission, and, if possible, the last dose should be given at least five hours before the expected chill. The follow-ing are some of the most useful anti-periodics, with the indications for their use :

Chininum sulphuricum.—This remedy is most useful when the paroxysms are well defined, especially in cases occurring in malarious districts. Headache and irrita-bility, in the first stage, are characteristic symptoms for the use of this drug. Give from five to twenty grains during the intermission.

Arsen.—Irregular paroxysms and great thirst during the first and second stages, anxiety and restlessness during the sweaty stage.

Eucalyptus globulus.—Well-defined paroxysms, with a long intermission and profuse sweats in the last stage. The dose is ten drops of the strongest tincture in sweetened water, repeated every two hours, during the intermission.

Ipecac. and *Nux* are sometimes indispensable remedies, in case of deranged digestion.

REMITTENT FEVER.

This is, essentially, a severe form of Intermittent Fever, in which there is no complete intermission of the fever between the paroxysms. In this form of malarial disease the febrile symptoms are more intense and the mortality is much greater than in ordinary Intermittent Fever. Remittent Fever is mostly confined to hot climates and pernicious malarious districts. The treatment is similar to that of Intermittent Fever, except that the antiperiodic remedies are given in divided doses and continued through the paroxysm.

TYPHOID FEVER.

DEFINITION.—A slightly contagious, slow fever, generated by putrifying animal matter.

SYMPTOMS.—For the first day or two, languor and uneasiness are almost the only noticeable symptoms. On the second or third day, slight chilliness, headache, loss of appetite, thirst, general weakness, and heat and restlessness at night. The tongue becomes brown and dry, and the pulse quick and soft. The feverish symptoms gradually increase until the seventh day, when there may often be found upon the abdomen a few rose-colored spots, about a tenth of an inch in diameter. The fever continues at its hight until the twenty-first day, after which there are marked morning remissions until the twenty-eighth day, by which time the fever has disappeared.

In severe cases, delirium sets in early and continues through the course of the fever. Diarrhœa is an almost constant symptom, and, in the later stages, the discharges frequently take place without the knowledge of the patient. There is, after the first week, some soreness of the abdomen, especially in the region of the right groin.

Among the bad symptoms are : swelling of the abdomen ; low, muttering delirium ; picking at the bedclothes; and sliding down in the bed. In the majority of fatal cases, death occurs about the twenty-first day.

Recovery is slow, and relapses may occur.

CAUSES.—The accidental leakage of sewers into wells of drinking-water; the stoppage of sewers, and consequent poisoning of the atmosphere in dwellings, by the gases which arise through the soil; badly ventilated or badly trapped water-closets. The above causes are much more potent in the production of Typhoid Fever than is daily contact with persons suffering from the disease.

PREVENTION.—Drains and sewers should be occasionally washed clean, by passing a strong current of water through them. If they are long, there should be a vent-hole at a distance from the house.

During the prevalence of the disease, all the alvine discharges from patients suffering from the fever should be received into vessels containing a strong solution of

6

Green Sulphate of Iron. Immediately after each passage, one or two ounces of *Strong Commercial Muriatic Acid* should be poured into the vessel. In cities, the contents of the vessel or bed-pan may be thrown into the water-closet. In the country, they may be thrown into trenches, dug for the purpose, in places where the drainage cannot be carried into drinking-water. It is safer, even then, to pour over the soil, after each deposit, a quantity of the *Muriatic Acid.*

TREATMENT.—The patient should take to the bed upon the first appearance of the disease. The sick-room should be dry and well ventilated. Carpets, bed-hangings and all unnecessary furniture should be removed. The services of a physician should be procured, if possible.

The daily use of the tepid sponge-bath will conduce greatly to the comfort and the recovery of the patient. Cold water should be freely administered as a beverage. The patient should be kept quiet and watched day and night, especially if there is active delirium.

Bell. is very useful in subduing the headache and delirium.

Bryon. is indicated if there is muscular soreness or swelling of the abdomen.

Arsen. is the sheet-anchor in the treatment of Typhoid Fever. The main indications are, constant thirst, dry tongue, diarrhœa, and great prostration.

DIET.—In a long and wasting disease like this, proper nutrition is absolutely essential to the recovery of the patient. All the food taken should be in a fluid or semi-fluid state. Of all foods, *milk* is the one best adapted to the needs of the fever patient. Many of the worst cases of Typhoid Fever have gone on to complete recovery with no medicinal treatment and no nourishment but milk.

Beef tea is very beneficial, after the active stage of the disease is passed. Its action is to stimulate the digestive powers and prepare the way for stronger food.

PREPARATION.—Take half a pound of clean, fresh beef-steak ; chop it in fine pieces, and place it in a wide-mouthed, glass bottle, holding a pint ; fill it with cold water, place it in a kettle, partly filled with cold water, and set it on the stove where the water will simmer for three or four hours. Then squeeze out and strain the liquid, salt it a little and set it aside to cool.

After the fever has left, the diet should be more nutritious. The following preparation of beef contains much more nutriment than *beef tea*. It may properly be called a *solution of beef*.

PREPARATION.—Take a pound of lean steak ; cut it into small pieces, and put it into a wide-mouthed, glass bottle ; add a pint of cold water, a teaspoonful of salt and twenty drops of *Chemically Pure Muriatic Acid* ; shake well and allow the mixture to stand, without heating, for two hours ; squeeze out the liquid.

The solution may be taken cold, or it may be gently heated before using.

As the digestive powers improve, jellies, and mutton or chicken broth, may be given. If the *beef tea* or the *solution of beef* is not relished by the patient, the yolk of an egg beaten with brandy is an excellent, stimulating food. Pounded raw beef and mixed diet may be allowed later. The greatest care should be taken not to allow the convalescing patient to glut himself, lest a relapse be brought on. Feed little and often.

GLANDULAR SWELLINGS.

The lymphatic glands of the neck or groins often swell and become tender to the touch.

CAUSES.—Taking cold ; inflammation of the throat ; Gonorrhœa ; Syphilis.

DIAGNOSIS.—The swollen glands are felt as little kernels or lumps which move freely under the skin. The swelling of the parotid glands in Mumps is close up under the ear and behind the jaw-bone. The lymphatic glands are scattered over various parts of the neck.

TREATMENT.—The remedies are : *Bell.* in acute, and *Merc. iod. rub.* in chronic cases.

GOITRE.

DEFINITION.—A chronic enlargement of the thyroid gland.

CAUSE.—The use of drinking-water containing magnesian limestone in solution. The frequency of the disease, in a country or section, varies with hardness of the water.

SYMPTOMS ETC.—The swelling is situated on the front portion of the neck, a little above the middle. It has a " doughy " feel. There is little or no pain or soreness. The tumor gradually increases till it becomes so large as to partially obstruct the breathing.

TREATMENT.—The patient should drink only soft water, or water which has been boiled. Give, internally, *Merc. iod. rub.* three times a day for several weeks. This treatment, if adopted early and used persistently, will often cure the disease.

If no benefit is seen in the course of four to eight weeks, an ointment, made by mixing forty-eight grains of finely pulverized *Iodide of Cadmium* with one ounce of *Cosmoline*, may be rubbed into the skin over the seat of the swelling. Use only a small portion at a time, and apply the ointment every other night.

GONORRHŒA.

DEFINITION.—A contagious inflammation of the mucous membrane of the genital organs.

SYMPTOMS.—For the first day or two, there is a burning sensation or a slight soreness felt in passing water.

These symptoms rapidly grow more intense, and there appears, on the second or third day, a discharge of a milky or creamy fluid, which increases in amount till by the end of the first week it amounts to several teaspoonfuls a day. The calls to urination are frequent and the pain is intense. This stage of the disease lasts from one to three weeks, at the end of which time the pain and discharge begin to diminish.

The third, or declining stage of the disease, is of variable duration. Much depends upon the treatment. The disease may be cured in a week, or it may continue for months or years, and materially shorten the life of the patient. The principal danger in the male subject is from the development of a "*stricture*," or a narrowing of the urethral canal.

CAUSE.—This disease is almost always caused by local contact with the specific poison, but in rare instances may be produced by leucorrhœal discharges.

TREATMENT.—The case should be, from beginning to end, under the direct, personal supervision of a regular, educated physician and surgeon. Shun all advertising specialists.

The internal use of *Acon.* will help to shorten the inflammatory stage. *Thuja occidentalis* is an excellent specific in the later stages. The dose is one drop of the strongest tincture, repeated every four hours.

GOUT.

DEFINITION.—An inflammatory disease caused by the accumulation of peculiar morbid matters in the blood, and characterized by pain and swelling of the first joint of the great toe.

CAUSES.—Luxurious living and indulgence in the use of alcoholic drinks.

SYMPTOMS.—The most prominent symptom is an acute torturing pain in some of the smaller joints, usually in the first joint of the great toe, sometimes in the heel, the ankle, the knee, the hand, the wrist, or the elbow. A fit of acute Gout is a succession of small fits, lasting, each, from three to twenty-four hours. Chronic Gout is accompanied by a deposit of chalky matters in and around the affected joint, which produces permanent stiffening. Sometimes the pain of Gout is suddenly transferred from the extremities to the stomach or bowels.

TREATMENT.—The affected part should be warmly encased in cotton or wool. The pain may be greatly relieved by the local application of a liniment composed of equal parts of strong tincture of *Aconite, Chloroform, Alcohol* and *Glycerine.*

The most successful internal remedy is *Colchicum autumnale*, drop doses of the strong tincture every half hour. If there is much thirst, coated tongue and other dyspeptic symptoms, give *Bryon.* If there is great constipation, give *Podoph.*, a dose every three hours. *Kali iod.* is a valuable constitutional remedy. It should be taken between the attacks. *Gels.* will afford relief in case of metastasis to the stomach or bowels.

DIET.—The food should be light. During the attack, no animal food or pastry should be eaten. Daily exercise in the open air and frequent bathing, will help to ward off future attacks.

GRAVEL.

DEFINITION.—A deposit of solid particles in the bladder.

A deposit which forms after urine has been passed, is called a sediment. A deposit which becomes so large and so solid that it cannot pass out of the bladder by a natural process, is called a stone. The latter sometimes forms in the kidney as well as in the bladder.

SYMPTOMS.—Frequent desire to urinate ; pain at the end of the act of passing water ; the passage of blood, mucus or gravel ; sudden stoppage of the stream, which starts again on a slight change of position. The presence of a stone in the bladder can best be determined by a surgeon.

TREATMENT.—Only soft water should be used for drinking. If necessary, the patient should change his residence, in order to get a supply of pure water. All excesses of diet and the use of intoxicating drinks, should be avoided. Lemonade and Carbonic Acid Water, ("Soda Water") help to dissolve the deposits as they form in the bladder. Water containing a little of the common Bicarbonate of Sodium, is sometimes useful for this purpose. For the relief of the spasm which accompanies the passage of gravel or stone, *Gels.* is a valuable remedy. The curative treatment is mainly preventive. Sometimes a stone has to be removed by a surgical operation.

GUM BOIL.

DEFINITION.—A small abscess in the socket of a tooth.

CAUSES.—Decayed teeth and exposure to cold.

SYMPTOMS, ETC.—There is a throbbing pain at and around the root of the tooth, and soreness both of the gum and of the tooth. The inflammatory process may or may not result in the formation of pus. When matter forms, it is usually discharged along the side of the tooth, but may make an opening through the cheek.

TREATMENT.—Apply hot fomentations to the cheek. Give *Acon.* and *Merc. viv.*, internally. If the tooth is too far gone to be saved by filling, have it extracted.

HAY FEVER.

DEFINITION.—A periodical Catarrh, affecting certain persons regularly during the summer months.

CAUSES. (A.) *Predisposing.*—A nervous organization, residence in a city, and overwork in literary or professional employments.

(B.) *Exciting.*—The breathing of air containing the pollen of certain plants, especially grasses. The pollen is the fertilizing-dust of the flower.

TREATMENT.—As yet no satisfactory medicinal treatment has been discovered. All the supposed discoveries have proved failures. The only known cure is to remove the patient to a locality where the exciting cause does not prevail. A sea voyage during the period of the disease, almost always gives immunity. Fire Island, near New York, the vicinity of the Straits of Mackinaw, and other localities, have been found to afford freedom from the annual attacks.

HEADACHE.

CAUSES.—Indigestion ; prolonged excitement of the nervous system ; the excessive use of Alcohol ; Catarrh ; Rheumatism ; Neuralgia ; Fevers ; and organic disease of the brain or its coverings.

REMEDIES. — *Acon.* — Violent, burning, compressive pain, especially below the root of the nose ; symptoms of fever.

Arnica—Headache caused by blows or other mechanical injuries ; cold feet.

Arsen.—Periodical headaches ; nausea, and bloated feeling in the stomach.

Bell.—Throbbing pain with sense of fullness over the eyes ; redness of the face.

Bryon.—Beating or shooting pains aggravated by movement ; irritable mood.

Chininum arsenicosum, second or third decimal trit.— Periodical headache with ringing in the ears; headaches from malarial poisoning.

Ignatia—Headache, temporarily relieved by change of position, sharp, boring pain, nervous exhaustion.

Nux—Pain in one side of the head; dull, stupid or dizzy feeling; sour stomach; constipation.

Physostigma venenosum, third decimal trit.,—Pressure in the temples with a constrictive sensation in the stomach, or a feeling as though there were a hard substance in the stomach.

HEART,—DISEASES OF THE.

DEFINITION.—Under this head are included only the real or organic diseases of the heart.

They may be divided into *acute* and *chronic*. The most common acute diseases of the heart are inflammation of the sac which encloses the heart (Pericarditis), and inflammation of the membrane which lines its cavities (Endocarditis). The chronic disorders of the heart are generally the results of Endocarditis. They are partial obstruction of the orifices (Valvular disease), imperfect closure of the valves (Valvular disease), and enlargement (Hypertrophy).

PERICARDITIS.

CAUSES.—Acute Rheumatism is the most common cause. It may however, come on insidiously during inflammation of the lungs, inflammation of the pleura, or inflammation of the kidneys. It is sometimes developed during the course of Small-pox and Scarlet Fever.

SYMPTOMS.—When it comes on in the course of Rheumatic Fever, there is pain in the region of the heart, and the breathing is oppressed. The pain is increased by pressure. The impulse of the heart is at first increased, but if water or pus is poured out into the sac, the impulse is diminished in intensity, though it may be felt over a greater surface. The patient is extremely anxious and restless. The disease runs its course in about ten days and recovery is nearly perfect. When the inflammation arises from causes other than Rheumatism, it is seldom recognized before the fatal event. Its course is rapid and the premonitory symptoms are generally absent.

TREATMENT.—*Acon.* and *Bryon.* are the most important remedies. Perfect quiet must be insisted upon.

ENDOCARDITIS.

CAUSES.—Rheumatism, Gout, Syphilis and the use of Alcohol.

SYMPTOMS.—Discomfort and uneasiness in the region of the heart, quick pulse, palpitation of the heart. The symptoms come on gradually and usually run into those of chronic valvular disease.

TREATMENT.—The remedies are *Acon.* and *Bryon.* *Kali iod.* has been recommended as a remedy capable of preventing the formation of the deposits which cause Valvular disease. It is held by some authorities to be an established fact that the free use of alkalies during Rheumatic Fever, will greatly diminish the danger of disease of the heart.

A teaspoonful of ordinary cooking soda, taken once in four hours, will keep the secretions alkaline, and not much less than this amount will answer the purpose.

VALVULAR DISEASE OF THE HEART.

CAUSE.—The deposit of fibrinous lymph on or about the valves of the heart. This often takes place as a result of Acute Rheumatism. Either of the four sets of valves of the heart may become incapable of perfect closure, and allow a backward flow (regurgitation) of blood at each contraction ; or one of the orifices may be permanently narrowed, so as not to allow a free, forward passage of blood.

SYMPTOMS.—Shortness of breath ; palpitation of the heart ; irregular beating ; occasional skipping of a beat ; dry cough ; choking sensation ; occasional attacks of dizziness or fainting ; swelling of the feet ; general dropsy ; blue skin ; inability to lie down without choking ; enlarged area of the place in which the beating of the heart may be felt ; a transfer of the point of greatest intensity upwards and to the left. The normal position of the "apex beat" is about an inch and a half to the right of, and below, the left nipple.

TREATMENT.—This form of disease is incurable. Great benefit has in many cases been derived from the use of *Cactus grandiflorus* and *Arsen.*, two or three doses of each daily. The second dilution of *Cactus* on pellets is the most convenient form for use. The patient must avoid all active exercise, running and lifting, and everything which tends to bring on the palpitation.

HEART,—PALPITATION OF THE.

CAUSES.—There are various causes of palpitation besides disease of the heart. Among the most common are : the formation of gases in the stomach in Dyspepsia ; worms ; constipation ; suppression of the menses ; thinness of the blood ; and troubles of the mind.

It is important to determine whether there is organic disease of the heart. This may be done by a physician who is skilled in "auscultation." One of the best signs of disease of the heart is that the worst spells or attacks of palpitation are brought on by *exercise only.* Palpitation from other causes comes on without warning, and may even wake the patient out of sleep.

TREATMENT.—This affection is curable in nearly every case, except those of valvular disease of the heart.

The cause should be carefully sought for and the treatment directed according to rational principles.

Among the best remedies are *Nux, Arsen.* and *Cactus grandiflorus. Acon., Ignat., Puls.* and *Santonin* are sometimes indicated. *Nux* and *Arsen.* are most useful in the dyspeptic form. *Ignat.* and *Puls.*, in case the cause is purely mental. *Cactus,* if there is irregular beating; and *Santonin,* if there are symptoms of worms.

HEMORRHAGE

DEFINITION.—This term is applied to almost every escape of blood from the body except that which takes place in the natural menstrual flow.

NOSE BLEED.

CAUSES.—This affection is much more common in youth than at any other period of life. Any occupation which requires stooping, tends to bring it on. Over-eating, and over-heating the head, are frequent causes.

TREATMENT.—Apply cold to the head and back of the neck ; keep the body in the upright posture and hold the arms above the head. If the bleeding is obstinate, it may be stopped by plugging the nostril with pieces of sponge which have been moistened with water and dipped in *Tannic Acid.*

BLEEDING FROM THE LUNGS.

(See " Consumption.")

VOMITING OF BLOOD.

CAUSES.—Ulceration of the mucous membrane, or of a cancerous growth in the stomach ; vicarious menstruation.

SYMPTOMS.—The vomited blood is thick and dark, and often looks like coffee-grounds.

TREATMENT.—Keep the patient quiet and give *Ipecac.*

HEMORRHAGE FROM THE BOWELS.

(See " Dysentery " and " Piles.")

HEMORRHAGE FROM THE UTERUS.

(See " Labor.")

HEMORRHAGE FROM WOUNDS.

TREATMENT.—The principal means of checking Hemorrhage from capillaries, veins and small arteries, are pressure applied directly to the wound, and the local use of cold water or ice.

In case of very profuse Hemorrhage, if a handkerchief be tied tightly around the arm or leg between the bleeding surface and the body, the flow of blood will be greatly diminished.

HICCOUGH.

CAUSES.—Swallowing air with the food; some forms of Dyspepsia.

TREATMENT.—Give the patient a few swallows of cold water or milk. If the affection is obstinate, give *Nux.*

HYDROPHOBIA.

DEFINITION.—A disease caused by a specific animal poison, and mostly propagated by the bites of dogs suffering from *Rabies.* The wolf, the jackall, the cat, the horse and the cow, have been known to have the disease.

SYMPTOMS.—*In the dog,* there may be first noticed, an uneasy, roving, shy, sullen manner The animal will refuse ordinary food, and go about eating straws and other rubbish. The sound of the bark is hoarse and unnatural. There is, usually, but not always, a disposition to bite at every person or animal in reach. There is no aversion to water, as is popularly supposed. The dog dies in from four to six days from the attack of the disease.

In man, the symptoms begin to develop in from a few weeks to two years from the time of receiving the bite of a mad dog. The disease begins with symptoms of fever. The first decided symptom is a spasmodic affection of the throat. The attempt to swallow, excites convulsions, and although the patient is tormented with thirst, he dares not drink, for fear of bringing on the convulsions. Great fear and excitement prevail throughout the course of the disease. The convulsions increase in severity, and at length come on without any exciting cause, and continue till the patient dies.

TREATMENT.—By far the largest proportion of biting dogs are not suffering from Rabies ; and the bites of a mad dog are not, in half of the cases, followed by Hydrophobia. Hence, it is readily seen that the wisest course is to confine the suspected dog, and watch him for a week or more, in order to determine whether he has Rabies. A terrible burden will be lifted from the minds of a bitten person and his friends when it is found that the dog was not mad. The wound should be seared, as soon as possible, with a white-hot iron or burned out with some strong caustic, such as strong *Carbolic Acid* or strong *Nitric Acid.* The hot iron is preferable, because it causes less pain, is more effectual and the wound heals more readily after it than after the acid caustics. " Mad_ stones " are relics of the age of superstition. Medicines are of no avail. The immediate and thorough cauterization of the wound is the only preventive of the disease. It is well to have it generally known that the virus is often wiped off the dog's teeth, as they go through the clothing. The danger from the bite of a mad dog is lessened by this means, but the clothing may become a source of infection.

HYPOCHONDRIASIS.

DEFINITION.—A chronic, functional disease of the nervous system, in which the patient imagines himself to be suffering from some dangerous organic disease.

CAUSES.—Nervous shocks; failure in business; idleness.

TREATMENT.—The patient's mind should be directed away from himself by getting him into some pleasant employment. Out-door life and change of scene and companionship, are more important than medication. In fact the administration of medicines may do harm, by fostering the attention to self, which is so important a feature of the disease.

Nux may be useful in relieving dyspeptic symptoms, and *Ignatia*, in stimulating the nervous system to healthy action.

HYSTERIA.

DEFINITION.—A functional disease of the nervous system, in which the patient feigns disease.

SYMPTOMS.—Hysteria may simulate almost any disease which is described in the books. The patient may seem to be suffering from Pleurisy, Heart Disease, Neuralgia, Rheumatism, Spinal Disease, inability to urinate, Epilepsy, or even to be in the last stages of Consumption. The imitations are so artful and so close as to deceive inexperienced bystanders.

The following peculiarities of the hysterical fit distinguish it from every other disease ; the attack always comes on when some one is present to witness it. If the patient falls, she contrives not to be hurt. The eyes are nearly closed, but the pupils may be found to follow

7

the movements of persons in the room. The breathing
is loud and irregular. The cough may occasionally stop
for a considerable time, or the lost voice be restored
while the patient is thrown off her guard by some temporary excitement. The fit often begins or ends with
laughter.

CAUSES.—Debility and nervous exhaustion.

TREATMENT.—The patient should be firmly encouraged
to think that she will soon recover. No sign of fear or
impatience should be shown. The name of the disease
should not be mentioned in her hearing.

Camphora, two drops of the strong tincture, on sugar,
every hour, will exercise a marked control over the
paroxysms, and in the majority of cases will be all the
medicine needed.

Ignatia is useful when the patient complains of a "ball
in the throat." *Puls.* or *Macrotin* will render assistance
if there is irregular or suppressed menstruation.

To prevent the attacks, the patient should take out-
door exercise, and avoid late hours and fatiguing mental
employments.

INFLAMMATION OF THE BLADDER.

DEFINITION.—Inflammation of the mucous membrane
of the bladder.

CAUSES.—Cold, damp feet ; holding the water too long;
gravel or stone in the bladder ; Gonorrhœa ; Stricture.

SYMPTOMS.—Sense of heaviness, with tenderness and
pain in the region of the bladder ; frequent urination ;
in chronic cases, the passage of blood or mucus with
the urine.

TREATMENT.—In acute cases, give *Acon.* or *Gels.* every hour. *Apis* or *Cantharis* may be required.

In chronic cases, *Merc. cor.* and *Sulphur* are the most reliable medicines. *Thuja occidentalis,* drop doses of the tincture, will help, in cases of great irritability of the bladder. Alcoholic drinks, especially gin, should be abstained from.

INFLAMMATION OF THE BOWELS.

DEFINITION.—Inflammation of the outer or serous coat of the contents of the abdomen (Peritonitis), or inflammation of the mucous and muscular layers of the intestines (Enteritis).

PERITONITIS.

CAUSES.—Surgical injuries ; extension of inflammation from the liver or intestine outward ; absorption of animal poison after childbirth.

SYMPTOMS.—Chills ; fever ; small, quick, hard pulse ; extreme tenderness of the abdomen, with pain at every breath.

TREATMENT.—Keep the patient quiet in bed.

Acon. and *Bryon.* are the best remedies. If there is headache or delirium, give *Bell.*

ENTERITIS.

CAUSES.—Taking cold ; errors in diet ; worms.

SYMPTOMS. — The symptoms differ from those of Peritonitis, in that the pain and soreness are not so extensive, being confined to the region of the navel. In Enteritis, likewise, there is less pain, less fever and usually some nausea and vomiting. In both forms of disease the bowels are constipated, and an attempt to move the bowels by cathartics might set up an inflammation, which would prove fatal to the patient.

TREATMENT.—*Acon.* is usually required at the beginning of the attack, and may, with advantage, be given throughout, if the febrile symptoms continue.

Arsen., if there are burning pains with thirst and distention of the abdomen.

Merc. cor., if there is straining and urging to stool, with passage of mucus. The patient should be kept quiet in bed, and allowed ice and cold water with a little milk or gruel. No solid food should be given.

INFLAMMATION OF THE BRAIN.

DEFINITION.—This includes, inflammation of the membranes surrounding the brain (Meningitis), and inflammation of the substance of the brain (Encephalitis). Meningitis is much more frequent.

MENINGITIS.

SYMPTOMS.— Chills ; hot skin ; quick, sharp pulse ; intolerance of light or noise ; sleeplessness ; delirium ; flushed face ; redness of the eyeballs ; vomiting, without nausea or retching. The dangerous symptoms are, convulsions, stupor, paralysis and coma.

CAUSES.—Measles ; Scarlet Fever ; Small-pox ; Typhoid Fever ; Rheumatism ; Tuberculosis ; blows upon the head; intemperance. Sometimes the disease appears as an epidemic, and comes on without any apparent cause. In these cases, the membranes covering the spinal cord, also, are sometimes affected, and the disease is called Epidemic Cerebro-Spinal Meningitis.

TREATMENT.—Keep the patient quiet, in a dark and well-ventilated room. Apply to the head, cold cloths or ice-bags (bladders or rubber bags filled with pounded ice).

Give, in the first stage, *Acon.* anu *Bell.;* in the latter stages, *Bell.* and *Helleborus niger,* third decimal dilution.

ENCEPHALITIS.

CAUSES.—Injuries to the brain ; the growth of tumors from the inner surface of the skull.

Softening of the brain is the result of a species of inflammation. One of the most frequent causes is intemperance.

SYMPTOMS.—The symptoms are much less marked than those of Meningitis. The febrile symptoms are comparatively slight. The pulse is weak and irregular. The face is not flushed. The breathing is irregular and sighing.

The peculiar symptoms of softening are : failure of memory ; confusion of ideas ; dimness of sight ; dullness of hearing ; irritability of temper : paralysis of one leg or of one side. The majority of patients with softening of the brain die within a month from the attack, though the disease may last from one to three years.

TREATMENT.—Rest, and nutritious food, are the most important means in the treatment. Fish and eggs should form a large portion of the food. *Acon., Bell.* and *Phos.* are the principal remedies. *Merc. viv.* will be found useful if there is evidence of Syphilitic taint.

INFLAMMATION OF THE BREASTS.

CAUSE.—The congestion incident to the starting of the secretion of milk. It is most common, immediately after the birth of the first child.

TREATMENT.—Give *Bell.*, internally, and rub the breasts thoroughly with *Arnica cerate*. *Belladonna cerate* is more certain and prompt in its action, but is hardly safe to use, if the patient is nursing a child, as it tends to dry up the milk, and it might prove injurious to the babe. If matter forms, a puncture should be made with a sharp, narrow-bladed knife. If a cut is made, the sharp edge of the knife should be turned toward the nipple.

INFLAMMATION OF THE EARS.

CAUSES.—Taking cold ; the extension of inflammation from the throat, as in Scarlet Fever or in Chronic Catarrh.

SYMPTOMS.—Buzzing noise in the affected ear ; pain in the ear ; headache ; dullness of hearing.

TREATMENT.—*Bell.* and *Puls.* are the remedies, in the first stage. *Merc. viv.* or *Silicea* should be given if there is a discharge. It is well in every case to put a plug of cotton in the ear. If the pain is acute, the cotton may be saturated with a mixture of one part of strong tincture of *Aconite* and four parts of *Sweet Oil*. If the pain is very severe, one part of *Chloroform* may be added to the mixture. The application of external warmth will help to relieve the pain and inflammation. A hot, roasted onion is a convenient and soothing application.

Chronic discharge from the ears should never be neglected. Permanent deafness often results from this suppurative process. While there is a discharge, the ear should be syringed daily, throwing in, very gently, a stream of warm Castile soap-suds.

INFLAMMATION OF THE EYES.

CAUSES.—The presence of foreign bodies ; the effects of a cold ; the action of poisons ; impurities of the blood.

SYMPTOMS.—The inflammation may affect the mucous membrane of the lids or of the eyeball, or even the deeper tissues, as the sclerotic coat and the iris.

In the common, catarrhal form, there is, at first, a feeling as though there were sand in the eye, followed by a watery secretion and redness of the blood vessels ; later, the eyelids swell and the secretions contain pus. If the deeper tissues are affected, there is intolerance of strong light. In the chronic form, which is sometimes called Scrofulous Ophthalmia, the lids remain red and swollen near the edges, and they are often glued together in the morning, by the drying of the secretions of the night.

TREATMENT.—The following remedies will be found most serviceable.

Acon.—Redness and swelling, in the acute stage ; intolerance of light.

Bell.—Deep-seated pain, with great intolerance of light and noise.

Kali iod.—Profuse watery secretion from the eyes and nose, with stuffed feeling in the head.

Merc. cor.—Chronic redness of the mucous membrane, with discharge of pus.

Merc. viv.—Feeling as though there were sand in the eyes ; pustules on the eye-balls or lids.

Puls.—Watering of the nose and eyes, with itching, burning or smarting.

Rhus.—Swelling, itching and soreness of the lids, with dryness of the eyes.

Sulphur.—Chronic cases, not cured by other remedies.

Local means. In acute cases, great benefit may be derived from the use of *Aconite,* locally. Prepare a mixture of one part of the strongest tincture and ten to twenty parts of clear, soft water, and lay on the eyes a light linen cloth moistened with the same. In chronic cases, where there is a gluey or scaly formation on the edges of the lids, they may be bathed every night with sweet milk. The oily substance in the milk, keeps the secretions moist till they can be washed off in the morning.

Many cases require to be kept in a darkened room, to prevent the irritation of strong light. The greatest care should be taken to avoid getting dust in the eyes, while they are inflamed.

INFLAMMATION OF THE KIDNEYS.

DEFINITION.—This includes, Acute Inflammation of the Kidneys (Acute Nephritis), and Chronic Inflammation and degeneration of the kidneys (Bright's Disease).

ACUTE NEPHRITIS.

SYMPTOMS.—Chilliness ; vomiting ; fever ; pain on each side of the spine, just above the hip-bones ; painless swelling of the feet, legs and other parts of the body. The urine thickens if it is boiled, showing the presence of albumen.

CAUSES.—Dr. G. Johnson has found, by an examination of two hundred cases, that fifty-eight were produced by the use of intoxicating drinks, fifty by exposure and twenty by Scarlet Fever.

TREATMENT.—Give the patient a warm bath and excite a perspiration as soon as possible, and keep the skin moist during the course of the disease. The remedies are :

Acon.—Chilliness ; fever ; thirst ; and scanty urine.

Arsen.—Dropsical swellings.

Merc. cor.—Mucus, blood or pus in the urine.

Apis mellifica, third decimal trit., may be given in alternation with *Arsen.*, for the relief of the dropsical symptoms.

CHRONIC NEPHRITIS.
(BRIGHT'S DISEASE.)

SYMPTOMS.—Gradually approaching debility ; pale, bloated appearance ; loss of appetite ; sometimes, but not always, dropsy. The urine is light in specific gravity (1.004 to 1.015), and forms a thick, white deposit of albumen when it is boiled.

CAUSES.—Hereditary tendency ; intemperance ; frequent exposure to cold ; Scarlet Fever ; Gout.

TREATMENT.—The secretions of the skin should be kept active by frequent hot-water, or hot-air baths. It is better, if convenient, that the patient remove to a warm climate, where an out-door life can be followed.

Arsen. and *Merc. cor.* are the principal remedies. The indications are the same as in Acute Nephritis.

Diet.—The food should be mainly vegetable. Milk has been found to be an excellent food in cases of Bright's Disease.

INFLAMMATION OF THE LIVER.

Definition.—Several varieties of Inflammation of the Liver have been distinguished, according to the tissue involved. There may be inflammation of the capsule, (Perihepatitis); of the glandules, (Hepatitis); of the connective tissue, (Cirrhosis); or of circumscribed masses, (Abscess).

Congestion of the Liver, the first stage of Hepatitis, is the most common of all derangements of this organ, and is the only one which properly comes within the limits of this work.

Symptoms.—Congestion of the Liver is usually accompanied by soreness, at the lower border of the ribs on the right side, and at the pit of the stomach, and pain in the back, under the right shoulder-blade. The countenance is sallow; the tongue, coated; the pulse, slow; the spirits are depressed; the bowels, constipated; and there is a general condition of laziness, with dull headache and dizziness on stooping.

Causes.—Overeating; too free use of fats and spirituous liquors; sedentary habits.

Treatment.—The diet should be abstemious, especially as regards fats and rich pastry. The patient should take daily exercise in the open air. Horse-back riding is especially useful.

The best internal remedies are *Podoph.* and *Merc. iod. rub.* *Nux* is often indicated by the dyspeptic symptoms.

If there is thirst and soreness of the bowels, with constipation, *Bryon.* is preferable.

INFLAMMATION OF THE LUNGS.
(PNEUMONIA.)

SYMPTOMS.—A chill, followed by fever ; hot, dry skin ;
thirst ; quick pulse ; rapid breathing ; dry cough ;
scanty, frothy, white, or rust-colored expectoration. In
most cases, the disease is complicated with Pleurisy, and,
as a consequence, every breath gives the patient a sharp
pain which causes him to press his hand on the affected
side, to relieve the distress. A characteristic sign of disease of the lungs, is that the breathing is quick in proportion to the pulse. The normal ratio is about four
beats of the pulse to one full respiration, (more accu-
rately, nine to two); in Pneumonia, there may be one
respiration to every three beats of the pulse, or, in
severe cases, one respiration to every two beats. The
severity of the disease varies, according to the amount
of lung tissue involved. Portions of both lungs may
be affected simultaneously, or one lung may be attacked
just as the other is recovering, or only one lung, or a
portion of one lung, may be affected. When the inflammation is at its hight, a dull, flat sound is heard
when the chest is struck over the inflamed and solidified
tissue, while other portions give a hollow, loud sound.

The duration of the disease is about three weeks, but
if there are relapses, it may be six weeks or more.

Broncho-Pneumonia, or Catarrhal Pneumonia, is a
form of inflammation distinct from that above described.

It comes on in the course of other diseases, such as Measles, Bronchitis, and Whooping-cough. The cough, at first loose, becomes gradually more oppressive and choking in character, and the patient suffers from want of air.

CAUSES.—Pneumonia proper, is rarely, if ever, caused by taking cold. The cause is obscure, but the disease generally comes on in a debilitated state of the system. Broncho-Pneumonia is caused by the progress of a bronchial inflammation downwards.

TREATMENT.—Fresh, cool air, and frequent cooling baths are most important adjuvants to the medicinal treatment. There is no danger of taking cold while the skin is hot with fever; but, of course, the strength of the patient must be taken into account, and the temperature and duration of the bath must be tempered according to its effect.

The indications for medicines are :

Acon.—In the first stage ; high fever and great restlessness.

Arsen.—In the advanced stages ; great anxiety ; failure of the vital forces ; coldness or blueness of the skin.

Bell.—Violent headache ; delirium ; starting during sleep.

Bryon. — Reddish, or rust-colored expectoration ; stitching pains in the sides.

Phos.—Frequent hacking cough, with tightness across the chest.

Tart. emet.—Broncho-Pneumonia ; accumulation of mucus which cannot be expectorated ; loose cough ; prostration.

INFLAMMATION OF THE TESTICLES.

CAUSES.—Mumps ; mechanical injuries ; the sudden suppression of a Gonorrhœal discharge.

SYMPTOMS.—The swollen testicle is hard, hot, and tender to the touch, and is usually quite painful.

A *Rupture*, in which the bowel slips down into the scrotum, might be mistaken for an inflamed testicle, but it may be distinguished by the fact that the rupture transmits an impulse to the fingers when the patient coughs, and slips back suddenly and entirely, with a gurgling noise, when it is pressed upward as the patient lies on his back.

Dropsy of the Scrotum may be distinguished by finding the testicle of the affected side as a small, movable body, distinct from the main swelling.

TREATMENT.—Put the patient in bed, and support the scrotum either by suspending it in a sling which passes up in front and is fastened to a band around the waist, or by propping it up from beneath. The pain, and even the swelling, are greatly relieved by the application of cloths wrung out of hot water.

The internal remedies are *Puls.* and *Merc. viv.* The latter will be needed only in case the swelling does not subside with the disappearance of the pain.

INFLAMMATION OF THE THROAT.

Definition.—This general tefm, includes inflammation of the upper and back portion of the throat (Pharyngitis), inflammation of the tonsils (Tonsillitis), and inflammation of the lower portion of the throat rear the windpipe (Laryngitis). Either two, or all of these forms of Sore Throat, may coexist.

PHARYNGITIS.

Causes.—Taking cold ; Dyspepsia ; specific poisons, such as Scarlet Fever and Syphilis.

Symptoms.—The inflamed portion of the throat feels raw, sore and dry ; or, in chronic cases, there may be a frequent clearing of the throat of a stringy mucus. The color of the mucous membrane may be bright red, dark red, or pale, with straggling, enlarged blood-vessels, or even ulcers, scattered over the surface.

Treatment.— *Bell.*— Acute cases ; dryness of the throat, with constant desire to swallow saliva.

Merc. viv.—Acute or chronic cases, where there is a tendency to ulceration.

Arsen.—Chronic inflammation, with dryness of the throat ; sour stomach ; bloated feeling in the stomach or abdomen.

TONSILLITIS.

Symptoms.—Chill ; fever ; general weakness ; soreness and pain in the throat, with frequent swallowing of saliva. The tonsils may be seen, like almonds, on each side of the throat. The inflammation may result in the formation of pus in one or both of the tonsils. The discharge is usually a good indication of approaching recovery.

CAUSES.—The same as in Pharyngitis. One attack seems to predispose the system to others.

TREATMENT.—*Acon.*—High fever.

Bell.—Dryness of the throat, with a deep red color.

Merc. iod. rub.—Great swelling of the tonsils ; a white or yellow coated tongue.

Arsen.—Slow recovery after suppuration ; dryness of the mouth and throat.

LARYNGITIS.

SYMPTOMS.—Loss of voice ; tickling sensation in the larynx.

CAUSES.—Same as Pharyngitis ; also, improper use of the voice.

TREATMENT.—Make no attempt to use the voice till the inflammation has subsided. Public speakers should avoid a prolonged use of the same pitch of voice.

Phos. is the remedy. In some chronic cases, *Hepar sulphuris calcareum*, sixth decimal trit., will afford great relief.

INFLAMMATION OF THE UTERUS.

SYMPTOMS.—There is, in acute cases, some fever, but in chronic cases, there is scarcely any. There is usually some pain, or burning and soreness, in the organ ; pain in the back and groin during stool ; pain in the rectum, extending forwards ; discharge of a thick, cheesy, or creamy substance, sometimes mixed with blood. In

chronic cases, the patient becomes nervous, and is subject
to Neuralgia and Hysteria.

In acute inflammation, rest alone is almost sufficient
to effect a cure, in good constitutions. *Acon.* and *Puls.*
will assist in removing the inflammatory action.

In chronic cases, the greatest pains must be taken to
improve the general health by baths, open-air life, and
a proper amount of rest. A change of residence for a
time will often be found beneficial.

Macrotin will accomplish marked beneficial results in
these cases. It should be given once or twice a day for
several months.

Nux and *Sulph.* may be required to remove the constipation of the bowels, which generally accompanies
and always aggravates this disorder.

ITCH.
(SCABIES.)

DEFINITION.—A contagious disease of the skin, caused
by a small insect of the spider family.

SYMPTOMS.—The tender skin between the fingers, on
the wrist, and on the sides, just above the hips, is the
most common site of the disease. Itching is the
earliest and most prominent symptom. There soon appears an eruption of small pimples, which is aggravated by scratching.

The insect (*Acarus Scabies*) measures from $\frac{1}{100}$ to $\frac{1}{30}$
of an inch in length, and from $\frac{1}{133}$ to $\frac{1}{75}$ of an inch in
its greatest breadth. It has eight legs, which are short
in proportion to the size of the body. The itching is
caused by the burrowing of the female insect, to obtain

a place to lay her eggs. The male remains always upon the surface of the skin. The "burrows," or nests, are little, crooked passages just beneath the surface of the skin. They vary in length from $\frac{1}{12}$ to $\frac{1}{4}$ of an inch, and are just large enough to allow the proprietor to turn around. When full they contain fourteen eggs, of which the female has laid one a day.

TREATMENT.—Wash and scrub the body with strong, warm soap suds, from crown to sole. Then rub into the skin, a lotion of *Hepar sulphuris kalinum*, strength one-tenth. This lotion may be procured at a homœopathic pharmacy.

PREPARATION.—Take half an ounce of fresh Lime, and an ounce of flowers of Sulphur ; mix them thoroughly in a mortar and melt the Sulphur, by gently heating and stirring the mixture; let it remain hot a minute or two; then pour on half a pint of water. Stir again and pour off and strain the liquid. Keep in a well-stoppered bottle.

This solution should be applied every night for a week, by which time the insects will all be dead. Put on clean under-clothes, and pass a hot flat-iron over the inside of the outer garments before they are put on.

ITCHING ERUPTIONS.

DIAGNOSIS.—These may be distinguished from Itch, by the absence of the little "burrows," and by their different location, as well as their changeable character.

TREATMENT.—In acute cases, with febrile action, give *Acon.* If the eruption is bright red, and contains little, watery vesicles, give *Rhus.*, third decimal dilution. If the disease has a chronic form, *Sulph.* is the remedy. It should be used once a day for some weeks. If the eruption is dry and scaly, without much itching, *Arsen.* is the proper remedy.

8

JAUNDICE.

SYMPTOMS.—Unnatural yellowness of the skin and whites of the eyes.

CAUSES.—Congestion or inflammation of the liver ; obstruction of the gall-duct. The latter affection is accompanied by such torturing pain, that the jaundice is unnoticed.

TREATMENT.—When it is caused by congestion of the liver, *Podoph.* and *Merc. iod. rub.*, two doses of each, daily, will effect a cure. If there is decided inflammatory action, *Acon.* and *Phos.* should be given.

LEUCORRHŒA.

DEFINITION.—An unnatural discharge of mucus or pus from the vagina.

SYMPTOMS.—The discharges vary in color from white to yellow and green, and may sometimes be mixed with blood. The patient is reduced in strength ; she becomes nervous and irritable or desponding, and complains of pain in the small of the back or in the hips.

CAUSES.—Leucorrhœa is very common for a few months before and for a few months after child-bearing. It may be produced by prolonged nursing, or by any other causes which depress the general strength.

TREATMENT.—The diet should be nutritious but un-stimulating. The patient should be in the open air as much as possible. *Macrotin* in many cases is the only medicine required. The vagina should be syringed, two or three times a day, with warm water, and in case the discharge does not cease, *Hydrastis*, tincture, may be added to the injection, in the proportion of one to twenty. If the discharge is fœtid or irritating, a solution of *Carbolic Acid*, in water, of the strength of one to two hundred, will be useful.

MEASLES.

DEFINITION.—A specific, contagious eruptive disease, characterized by a catarrhal fever, which lasts from three to six days, followed by an eruption of small, red, slightly raised spots, which are arranged in curves.

SYMPTOMS.—In the first stage: chills ; headache ; backache ; quick pulse ; watering of the eyes ; over sensitiveness to light ; sneezing and coughing. The primary fever is much lighter than that of Small-Pox or Scarlet Fever.

The appearance of the eruption is followed by a great decline of the fever. The eruption is pinkish or raspberry-red, or in bad cases, dark purple, with a mixture of yellow or even black. The spots are distinct and are not larger than a flea-bite.

The most common, as well as the most serious, after effects of the disease, are Inflammation of the Lungs and Inflammation of the Eyes.

CAUSES.—Children are most susceptible to the disease, but no age is exempt. The poison spreads with the greatest facility ; no susceptible person can enter a room in which there is a patient with Measles, without contracting the disease.

TREATMENT.—Keep the patient in a room, moderately and uniformly warm. The bathing which is both safe and salutary in Scarlet Fever and Typhoid Fever, is dangerous in Measles. It is a common but grave error to give hot drinks in Measles. There is no surer way to bring on congestion and inflammation of the lungs. Give frequent cool drinks, but do not let a cold current of air touch the skin.

The remedies are *Acon.* and *Puls.* If there are symptoms of inflammation of the lungs, *Phos.* and *Bry.* will be useful. *Bell.* is a potent remedy to relieve the secondary Inflammation of the Eyes.

MENSTRUATION,—IRREGULARITIES OF.
SUPPRESSION OF THE MENSES.
(AMENORRHŒA.)

TREATMENT.—In case of sudden suppression, produced by exposure to cold or wet, give one or two doses of *Gels.*, and let the patient take a hot hip-bath or a hot foot-bath of about twenty minutes' duration.

If there is a tardy appearance of the menses, with paleness and general weakness, *Ferrum phosphoricum*, first decimal trit., two or three grains after each meal, for several weeks, will improve the condition of the blood, and thus bring about the natural flow. The patient should at the same time, be encouraged in outdoor exercises and amusements. If there are wandering pains, palpitation of the heart, or flushed face with headache, *Puls.* will often give quick relief. If there is a persistent pain in the side or back, or a dragging sensation in the lower portion of the bowels, give *Macrotin.*

PROFUSE MENSTRUATION.
(MENORRHAGIA.)

TREATMENT.—Keep the patient quiet during the period of flowing.

If the discharge is dark colored, give *Hamamelis* in drop doses every hour.

If the discharge consists of bright blood, give *Ipecac*. If the flow is copious, the application of cold cloths to the abdomen and thighs, may be necessary.

If the patient is full blooded and inclined to be feverish, give *Acon.* and *Bell.*

The weakness caused by the loss of blood may require, in acute cases, *Camphora*, or in the intervals of chronic cases, *Ferrum* as above.

If there is irregular menstruation with constipation, *Sulphur* is an excellent remedy.

PAINFUL MENSTRUATION.
(DYSMENORRHŒA.)

TREATMENT.—In acute cases, with some feverish symptoms (Spasmodic Dysmenorrhœa), *Gels.*, in five drop doses, every two hours, will often give great relief.

When the pain occurs just before or during the beginning of the flow, the probability is that there is congestion of the uterus (Congestive Dysmenorrhœa). In such cases give a dose of *Macrotin* every night, for a week preceding each menstrual period.

Painful menstruation may be caused by tumors or displacements of the uterus (Obstructive Dysmenorrhœa). If the remedies mentioned above, fail to give relief, the patient should consult a physician.

CESSATION OF THE MENSES.
(MENOPAUSIS.)

In the majority of cases, no medicine is required.

The nervous symptoms yield best to *Puls.* Constipation and Piles require *Nux* and *Sulphur.*

MILK CRUST.
(CRUSTA LACTEA.)

DEFINITION.—A variety of *Eczema,* which develops on the scalp of teething infants.

SYMPTOMS.—There is first, redness of a small portion of the scalp ; afterwards, there are small pimples which burst and exude a watery or purulent fluid, forming extensive crusts. The disease may be confined to a portion of the top of the head, or it may extend down the neck and cover the chest. There is considerable itching, and often some fever. The disease sometimes continues throughout the whole period of teething.

TREATMENT.—Wash the affected part daily with warm, soft water and pure Castile soap, and dry carefully by pressing with a soft cloth. If soft water cannot be easily obtained, boiled, hard water, to which has been added a little wheat bran, forms a good substitute. The application of a very small quantity of *Cosmoline* after each washing will help to prevent the formation of crusts. If the child scratches so hard as to draw blood, its hands should be covered with mittens when it cannot be watched.

The most useful remedies are :

Arsen., if there is vomiting ; if the countenance is pale ; if the eruption is dry ; or if the discharges are watery and irritating.

Sulph., in chronic cases, with violent itching.

Merc. viv., if there is a profuse oozing of pus, or if there is swelling of the glands.

Rhus, if there is itching and redness, with considerable watery discharges.

Lappa major, in drop doses of the strong tincture, three times a day, has been found curative in cases which resist other treatment.

Viola tricolor, first dilution, may be given in the same way if there are convulsions, hiccough, or other signs of affection of the brain.

Local applications which tend to dry up the eruption, are liable to produce dangerous internal diseases.

MUMPS.

DEFINITION.—A specific, contagious disease, characterized by inflammation of the parotid glands.

SYMPTOMS.—The disease is most common among children. There is a slight fever, followed by pain, heat, redness, swelling and soreness, just under the ear and behind the descending portion of the jaw. The pain is felt most when the patient attempts to open the mouth. Both sides may be affected simultaneously or successively, or only one may be attacked. In severe cases, the tonsils and the other salivary glands may be inflamed. In males the testicles, and in females the breasts, are sometimes attacked with inflammation, when the disease is at its hight.

TREATMENT.—In mild cases, it is sufficient to keep the patient in a warm room until the inflammation has ceased (about eight days). In severe cases, *Bell.* and

Merc. iod. rub. will help to reduce the swelling.   If the testicles or breasts are affected, give *Puls.*   The occasional application of cloths wrung out of hot water will relieve the soreness and pain.   If these applications are made, the swollen parts should be kept covered in the intervals with dry cloths.

NAUSEA.

CAUSES.—Eating improper food ; over-eating ; Dyspepsia ; pregnancy ; inflammation of the kidneys ; obstruction of the bowels ; ulceration of the stomach ; Scarlet Fever ; Small-pox ; Intermittent Fever.

TREATMENT.—In case of over-eating or eating improper food, vomiting should be encouraged till the offending substances are expelled.   The swallowing of warm drinks or tickling the throat with the finger will promote vomiting.

The remedies to relieve nausea and vomiting are :

Ipecac., for simple nausea or vomiting of slimy, greenish or blackish substances.

Arsen., for vomiting of watery fluids, followed by thirst ; burning in the stomach ; cold hands and feet.

Iris versicolor, third decimal dilution, in cases of bilious sick-headache.

Nux, for nausea, with colic or constipation.

NERVOUSNESS.

CAUSES.—Dyspepsia ; disease of the heart ; disease of the uterus ; overwork ; watching ; mental worry.

TREATMENT.—If there are dyspeptic symptoms, give *Nux.*   If there is fever, give *Acon.*   If the cause is overwork or prolonged nervous excitement, a cup of green tea will give temporary relief. If the trouble is caused by grief, *Ignatia* is the remedy. If there are symptoms of uterine irritation, *Puls.* or *Macrotin* will be found useful.

NETTLE-RASH.
(URTICARIA.)

DEFINITION.—A non-contagious affection of the skin, characterized by an eruption of raised spots with heat, burning and itching.

SYMPTOMS.—The spots resemble those produced by the stings of nettles or of mosquitoes. They are usually whitish in the centre, and red at the border, and vary from a twelfth to a half an inch in diameter. They contain no fluid. The eruption may be brought out in a new part, by scratching.

CAUSES.—Indigestion from eating particular kinds of food, such as bitter almonds, cucumbers, mushrooms, oatmeal, buckwheat flour, or shellfish; suppressed menstruation, teething, or the wearing of flannel.

TREATMENT.—*Acon.* and *Rhus.* are the most effectual, remedies. A general tepid bath, not followed by rubbing, will help to relieve the irritation. If flannel is worn, it should not be next to the skin. The diet should be light, simple and free from substances supposed to produce the disorder.

NEURALGIA.

DEFINITION.— Pain produced by functional disorder of some particular nerve.

SYMPTOMS.—The pain is sharp and darting and often intermittent in its character. The nerves most frequently affected are those of the face and jaws (fifth pair), and the nerves which pass down the outer and back portion of the thigh (Sciatic nerves), but almost any nerve of sense in the whole body may be affected ; as for instance, the nerves between the ribs (Intercostals), the nerves of the forehead (in Brow Ague), the nerves of the stomach, ovaries or testicles.

CAUSES.—Hereditary predisposition ; impairment of the general health ; thinness of the blood ; malaria ; exposure to cold drafts ; decayed teeth.

TREATMENT.—If the disease is caused by exposure to cold, the affected part should be kept warmly covered. Hot fomentations are generally beneficial.

The remedies are :

Acon., if there is fever, heat or swelling.

Arsen., if there are intervals of complete or partial relief ; or, if there is malarial poisoning.

Gels., if there is nervous or febrile excitement.

Puls., if the patient is extremely sensitive and the pain is transient in character.

POISONING.

GENERAL TREATMENT.—If a poisonous substance has been swallowed, introduce into the stomach, as soon as possible, the whites of several eggs, beaten up with water, or a large quantity of milk, and then give an emetic. One of the most convenient and effective emetics, is a tablespoonful of ground *Mustard*. A half teaspoonful to a teaspoonful of *Sulphate of Zinc*, is a quick-acting and safe emetic. After vomiting has commenced, more of the eggs or milk may be given, and the process kept up till the vomited matters seem to contain no more of the poison. In many cases it will be sufficient to tickle the throat with the finger or with a feather. No warm drinks should be given.

An antidote should be administered as soon as possible. The following brief list of poisons and antidotes will cover nine-tenths of all cases of poisoning:

ACID POISONS.—These are *Carbolic Acid*, *Muriatic Acid*, *Nitric Acid*, *Oxalic Acid*, *Prussic Acid* and *Sulphuric Acid*, etc. The antidotes are the Alkalies. Among the safest and surest are : common *Cooking Soda* (Bicarbonate of Sodium), dissolved in water ; *lime water* or *slaked lime* ; *pounded chalk* ; *pounded marble* and *Magnesia*. In default of any of these, scrape the *whitewash* from the walls and mix it with milk or white of eggs.

ALKALINE POISONS.—These are *Ammonia*, *Potash* (lye), *Soda* and *Lime*. The antidotes are the mild vegetable acids. *Vinegar* (Acetic Acid) is the most convenient as well as one of the most effectual. *Lemon juice* and *Tartaric Acid* are also good.

ARSENIC.—The symptoms produced by large doses of *Arsenic* are : nausea ; vomiting ; thirst ; burning in the stomach ; coldness of the skin ; diarrhœa ; bloating of the stomach and abdomen, or of the whole body ; difficulty of breathing. The larger the dose the sooner the symptoms appear and the more likely the patient is to throw

out the poison by vomiting. The antidotes are : a mixture of *Muriated tincture* of *Iron* and *Ammonia* (Hydra-
ted Peroxide of Iron) ; a mixture of *Magnesia* (Carbonate of Magnesium) and *Sugar; Milk of Magnesia* (Hydrate
of Magnesium).

ALCOHOL.—For the headache and indigestion following
a spree, *Nux* is a valuable remedy.

The stupor of Alcoholic intoxication, calls for cold applications to the head and occasional rousing.   Vomiting
should be encouraged.   It is often difficult to distinguish
between Alcohol poisoning, Opium poisoning, Apoplexy,
Epilepsy, Sunstroke, the effects of blows upon the head,
and Hysteria.   The following diagnostic symptoms may
be of service.   In Alcohol poisoning, the face is flushed
and an odor of alcoholic drinks may be detected in the
breath.   In Opium poisoning, the face is pale, the pul-
pils contracted and the breathing slow.   In Apoplexy,
there is paralysis, generally of the whole body.   In
Epilepsy,there are convulsions and frothing at the mouth.
In Sunstroke, the head and body are hot and the breathing is rapid.   In concussion of the brain, there are usually
external bruises.   In Hysteria, the patient may be
aroused by a proposal to pour boiling water on the face.

LEAD.—The *symptoms* of poisoning by lead come on
slowly.   They are : blueness of the gums, next to the
teeth ; pale face : occasional colic ; inability to raise the
hand, when it is held out with the palm downward ; in
protracted cases, general paralysis.

The means by which Lead is introduced into the circulation are various.   Painters become poisoned by eating with unwashed hands, or by breathing an atmosphere

loaded with the vapors of Turpentine, which convey small particles of Lead into the mouth, throat and lungs. Fruits preserved in tin cans, food cooked in glazed vessels, pickles kept in glazed earthenware, and soft water conveyed in lead pipes, are media through which lead is taken into the body. Hair dyes and face powders almost all contain this poisonous metal, and are frequent causes of obscure, wasting disease.

The treatment of Lead Colic consists in mitigating the pain by the use of hot baths (or, in responsible hands, the use of opiates) and removing the poison from the intestines by means of frequent purgative doses of Epsom Salts. The dose is one drachm.

The treatment of Lead Paralysis consists in removing the poison from the blood by means of substances which dissolve it so that it may be separated and carried off by the kidneys. *Kali bromatum*, in doses of five to ten grains, every night, will effect this end.

CHLOROFORM.—In case of threatened death from Chloroform, open the doors and windows, loosen the clothing about the neck and trunk, dash cold water in the face and imitate the motions of breathing by pressing in the ribs and letting them expand suddenly. If the breathing is obstructed, by the falling back of the tongue, turn the patient on his side. An open bottle of *Aqua Ammoniæ* may be held near the nose for a moment, but not longer.

MERCURY.—The acute poisonous effects of Mercury are : soreness of the gums ; loosening of the teeth ; spitting large quantities of saliva ; mucous or bloody diarrhœa. In chronic poisoning the teeth loosen and fall out, the jaw bones decay, ulcers form in the throat and various parts of the body, and the general health fails.

For the ulcerations in the mouth and throat a gargle of *Chlorate of Potassium*, thirty grains to the ounce, may be used with good effect. One of the best remedies for the general condition is *Nitri Acidum*, first decimal dilution in water, a dose of one drop between meals.

Opium, Morphine, Laudanum, Paregoric, Soothing Syrup, etc.—The stupor produced by an overdose of an opiate may be broken by frequent rousing and administration of drop doses of tincture of *Camphor* on sugar. The nervous irritability produced by the continued use of opiates is best relieved by coffee, or by *Bell.* and *Nux.*

Nitrate of Silver.—The antidote is common salt.

Phosphorus.—After the stomach has been emptied of the poison as much as possible by vomiting, give the patient *Magnesia* (Carbonate of Magnesium) or *Lime-water*, and considerable quantities of gruel, but on no account give any substance which contains much oil.

PILES.

Definition.—Enlargement of the blood vessels of the lower portion of the bowels, with thickening of the mucous membrane, causing hard tumors, either outside or inside the cavity of the rectum.

Symptoms.—Feeling of weight, itching and smarting at the lower end of the bowels. When the tumors are internal they are more likely to bleed.

CAUSES.—Sedentary habits, the use of fine foods, chronic constipation, pregnancy.

TREATMENT.—Remove the constipation by appropriate diet and medicines, and apply to the piles a cerate of *Æsculus.* If there is profuse bleeding, give *Hamamelis* tincture in five drop doses, and inject cold water.

PLEURISY.

DEFINITION.—An inflammation of the pleura (or membrane which lies between the lungs and the inner walls of the chest.)

In health the pleura has a smooth, lubricated surface, which allows free motion of the lungs in the cavity of the chest. In Pleurisy the smoothness of this surface is destroyed, and movements are painful.

SYMPTOMS.—There is almost always a chill, followed by considerable fever, with a sharp stitching pain in the side whenever the patient takes a deep breath or coughs.

CAUSES.—Taking cold, extension of inflammation from the lungs, mechanical injuries, such as a fracture of the ribs.

TREATMENT.—In the first stage, *Acon.* and *Bryon.* are invaluable remedies. In the later stages, when effusion has taken place between the lungs and the chest walls, *Merc. iod. rub.* is useful in stimulating absorption.

The application of heat to the seat of the pain will often afford relief. The application of long strips of adhesive plaster in various directions over the affected side in such a manner as to prevent a full expansion of the chest, will ease the patient and promote the cure.

PREGNANCY,
INCLUDING THE TREATMENT OF LABOR.

The signs of pregnancy are :

1. *Non-appearance of the menses* during the whole period of gestation. In exceptional cases there is a slight flow at the time of the first or second monthly period.

2. *Nausea and vomiting,* commonly called " Morning Sickness." This usually begins in the first month and continues till the end of the fifth month.

3. *Enlargement and tenderness of the breasts,* beginning in the first month and increasing till the fourth, and subsiding again during the remainder of the period.

4. *Changes in the areolæ of the breasts,* most noticeable during the first pregnancy. By the end of the second month, there is a slight darkening of the circle around the nipples, which increases till the end of the pregnancy. Little raised spots appear in the circle during the fourth month and increase till the ninth month.

5. *Enlargement of the abdomen,* beginning early in the third month, and increasing till the middle of the ninth month, after which there is a slight diminution in size. The upper border of the uterus reaches the navel by the middle of the sixth month.

6. *Changes of the navel.* During the first two months there is a slight sinking in of the navel. By the fifth month there is no depression, and during the remainder of the period there is generally a prominence of the navel.

7. The appearance of a scum on the urine after it has stood for several hours. This phenomenon begins in the third month and continues throughout the pregnancy.

8. *Light spots on the abdomen*, after the sixth month.

9. *Movements of the child.* These are first felt about the middle of the fifth month, and increase in strength and frequency till the period of labor.

10. *Flow of milk from the breasts*, during the third and ninth months.

11. During the eighth and ninth months, the following troublesome train of symptoms often arises : Constipation, difficulty of breathing, difficulty in walking, swelling of the feet and legs, itching of the lower limbs and lower part of the body, piles, leucorrhœa. The difficulty of breathing is relieved during the last two weeks by the descent of the uterus.

12. *The beating of the child's heart* may be heard, by placing the ear on the abdomen in the proper locality, from the beginning of the sixth month to the end of the ninth month.

TREATMENT OF THE COMMON DISORDERS OF PREGNANCY. —The " Morning Sickness " of pregnancy may be relieved, or even cured, by a few doses of the appropriate medicine. *Nux* and *Ipecac* are the most useful remedies. If there is heartburn and waterbrash or vomiting of watery fluids, with burning thirst, *Arsen.* will give prompt relief. Ice-cream, iced milk, soda-water (water charged with Carbonic Acid gas), jellies, animal broths, or the yolk of eggs, may be retained when ordinary food and drink is rejected. In bad cases, milk, with the addition of one-half its bulk of transparent lime water, will do excellent service by sustaining the strength, till other food can be digested. It is a good rule in such cases, to eat little and often.

The swelling of the lower extremities may be relieved by the use of *Hamamelis*, in drop doses, internally, and as a lotion, locally, aided by the use of elastic stockings. The patient should have the feet raised when sitting.

The troublesome itching may be allayed by the use of *Sulphur*, or by tepid sponge-baths.

It is well in all cases to toughen the nipples during the latter months of pregnancy, by bathing them frequently with *Alcohol*, or with a solution of *Tannin*, in water. Much suffering and annoyance may be prevented by this means.

If *hemorrhage* occur during pregnancy, or labor, the advice of a physician should be obtained without delay. Perfect rest should be enjoined, with the head low, and the feet and hips elevated.

Labor, or childbirth, comes on two hundred and seventy-three days after conception, or in the great majority of cases, about two hundred and eighty days after the beginning of the last menstrual flow. The following table, showing two hundred and eighty days from any given date, will be convenient for reference, especially to the young, expectant mother :

Jan. 1	Oct. 8	Jan. 19	Oct. 26	Feb. 6	Nov. 13	Feb. 24	Dec. 1
" 2	" 9	" 20	" 27	" 7	" 14	" 25	" 2
" 3	" 10	" 21	" 28	" 8	" 15	" 26	" 3
" 4	" 11	" 22	" 29	" 9	" 16	" 27	" 4
" 5	" 12	" 23	" 30	" 10	" 17	" 28	" 5
" 6	" 13	" 24	" 31	" 11	" 18	Mar. 1	Dec. 6
" 7	" 14	" 25	Nov. 1	" 12	" 19	" 2	" 7
" 8	" 15	" 26	" 2	" 13	" 20	" 3	" 8
" 9	" 16	" 27	" 3	" 14	" 21	" 4	" 9
" 10	" 17	" 28	" 4	" 15	" 22	" 5	" 10
" 11	" 18	" 29	" 5	" 16	" 23	" 6	" 11
" 12	" 19	" 30	" 6	" 17	" 24	" 7	" 12
" 13	" 20	" 31	" 7	" 18	" 25	" 8	" 13
" 14	" 21	Feb. 1	" 8	" 19	" 26	" 9	" 14
" 15	" 22	" 2	" 9	" 20	" 27	" 10	" 15
" 16	" 23	" 3	" 10	" 21	" 28	" 11	" 16
" 17	" 24	" 4	" 11	" 22	" 29	" 12	" 17
" 18	" 25	" 5	" 12	" 23	" 30	" 13	" 18

Mar.14	Dec.19	May17	Feb.21	J'ly20	Ap'l 26	Sep.22	June 29
"15	"20	"18	"22	"21	"27	"23	"30
"16	"21	"19	"23	"22	"28	"24	July 1
"17	"22	"20	"24	"23	"29	"25	"2
"18	"23	"21	"25	"24	"30	"26	"3
"19	"24	"22	"26	"25	May 1	"27	"4
"20	"25	"23	"27	"26	"2	"28	"5
"21	"26	"24	"28	"27	"3	"29	"6
"22	"27	"25	Mar. 1	"28	"4	"30	"7
"23	"28	"26	"2	"29	"5	Oct. 1	"8
"24	"29	"27	"3	"30	"6	"2	"9
"25	"30	"28	"4	"31	"7	"3	"10
"26	"31	"29	"5	Aug.1	"8	"4	"11
"27	Jan. 1	"30	"6	"2	"9	"5	"12
"28	"2	"31	"7	"3	"10	"6	"13
"29	"3	June1	"8	"4	"11	"7	"14
"30	"4	"2	"9	"5	"12	"8	"15
"31	"5	"3	"10	"6	"13	"9	"16
Apr.1	"6	"4	"11	"7	"14	"10	"17
"2	"7	"5	"12	"8	"15	"11	"18
"3	"8	"6	"13	"9	"16	"12	"19
"4	"9	"7	"14	"10	"17	"13	"20
"5	"10	"8	"15	"11	"18	"14	"21
"6	"11	"9	"16	"12	"19	"15	"22
"7	"12	"10	"17	"13	"20	"16	"23
"8	"13	"11	"18	"14	"21	"17	"24
"9	"14	"12	"19	"15	"22	"18	"25
"10	"15	"13	"20	"16	"23	"19	"26
"11	"16	"14	"21	"17	"24	"20	"27
"12	"17	"15	"22	"18	"25	"21	"28
"13	"18	"16	"23	"19	"26	"22	"29
"14	"19	"17	"24	"20	"27	"23	"30
"15	"20	"18	"25	"21	"28	"24	"31
"16	"21	"19	"26	"22	"29	"25	Aug. 1
"17	"22	"20	"27	"23	"30	"26	"2
"18	"23	"21	"28	"24	"31	"27	"3
"19	"24	"22	"29	"25	June 1	"28	"4
"20	"25	"23	"30	"26	"2	"29	"5
"21	"26	"24	"31	"27	"3	"30	"6
"22	"27	"25	April 1	"28	"4	"31	"7
"23	"28	"26	"2	"29	"5	Nov.1	"8
"24	"29	"27	"3	"30	"6	"2	"9
"25	"30	"28	"4	"31	"7	"3	"10
"26	"31	"29	"5	Sept.1	"8	"4	"11
"27	Feb. 1	"30	"6	"2	"9	"5	"12
"28	"2	July1	"7	"3	"10	"6	"13
"29	"3	"2	"8	"4	"11	"7	"14
"30	"4	"3	"9	"5	"12	"8	"15
May 1	"5	"4	"10	"6	"13	"9	"16
"2	"6	"5	"11	"7	"14	"10	"17
"3	"7	"6	"12	"8	"15	"11	"18
"4	"8	"7	"13	"9	"16	"12	"19
"5	"9	"8	"14	"10	"17	"13	"20
"6	"10	"9	"15	"11	"18	"14	"21
"7	"11	"10	"16	"12	"19	"15	"22
"8	"12	"11	"17	"13	"20	"16	"23
"9	"13	"12	"18	"14	"21	"17	"24
"10	"14	"13	"19	"15	"22	"18	"25
"11	"15	"14	"20	"16	"23	"19	"26
"12	"16	"15	"21	"17	"24	"20	"27
"13	"17	"16	"22	"18	"25	"21	"28
"14	"18	"17	"23	"19	"26	"22	"29
"15	"19	"18	"24	"20	"27	"23	"30
"16	"20	"19	"25	"21	"28	"24	"31

Nov.25	....Sept. 1	Dec. 5	.....Sept. 11	Dec.14	.....Sept. 20	Dec.23	.....Sept. 29
" 26.....	" 2	" 6.....	" 12	" 15.....	" 21	" 24.....	" 30
" 27.....	" 3	" 7.....	" 13	" 16.....	" 22	" 25.....	Oct. 1
" 28.....	" 4	" 8.....	" 14	" 17.....	" 23	" 26. ..	" 2
" 29.....	" 5	" 9.....	" 15	" 18.....	" 24	" 27.....	" 3
" 30.....	" 6	" 10.....	" 16	" 19.....	" 25	" 28.....	" 4
Dec. 1. ...	" 7	" 11.....	" 17	" 20	" 26	" 29.....	" 5
" 2.....	" 8	" 12.....	" 18	" 21.....	" 27	" 30.....	" 6
" 3.....	" 9	" 13.....	" 19	" 22.....	" 28	" 31.....	" 7
" 4.....	" 10						

The signs of the near approach of labor are : a slight diminution of the size of the abdomen, for some days previous; excitement; pains in various parts of the body; frequent desire to pass water, for some hours previous; a discharge of mucus, tinged with blood, for a very short time previous to the actual beginning of labor. The accession of labor is indicated by the occurrence of pains at regular intervals, at first an hour or more apart, but gradually increasing in frequency, till at the end of the labor there is not more than a minute between the pains. It is best to call the medical attendant as soon as the discharge of bloody mucus begins, or as soon as the pains come at regular intervals.

During the first stage of labor, if the pains are too frequent and too feeble, while the patient is nervous or feverish, a dose of five to ten drops of *Gels.*, in a cup of hot water, will quiet the fever, facilitate the dilatation of the uterus and hasten the labor.

If flooding occur, after birth of the child, "knead" the abdomen with the hands till a pain comes on, or sprinkle or pour cold water on the abdomen.

If the child does not breathe or cry almost immediately after its birth, it may require to be stimulated by sprinkling cold water on the face and breast. If this does not prove successful, one or more of the following measures may be tried : Slap the buttocks with the hand or a wet towel ; apply vinegar to the lips on a sponge or a cloth ; turn the child on its side, with the face a little downward, and clear the mouth of mucus by introducing a finger ; press the chest in such a manner as to imitate the motions of breathing ; inflate the chest by forcing air into the nose and mouth simultaneously. Avoid chilling the babe during this process.

Apply the child to the breast, as soon as it is dressed. On no account administer any medicine, food or drink during the first twelve hours. The mother's milk is all-sufficient. In case the secretion of milk is slow in starting, and the child seems hungry, it may be fed a little warm, sweetened water. If the milk does not start within twenty-four hours, the child may be fed occasionally a little warm cow's milk, with the addition of a teaspoonful of sugar to the teacupful.

After-pains may be relieved by giving *Secale cornutum*, strong tincture, in drop doses, every ten minutes. A folded towel, placed on the lower part of the abdomen, beneath the binder, will help to prevent the recurrence of the pains.

The beginning of the secretion of milk is sometimes accompanied by febrile action,—*Milk Fever*. The remedies are *Acon.* and *Bell.*

Sore nipples may almost always be prevented by proper attention to the treatment recommended in the former part of this article. During the period of nursing, it is well to wash the nipples with tepid water, just after each time the child has taken the breast ; dry the nipples carefully, and then dust over them some fine wheat flour. If excoriation or ulceration has taken place, apply, after the child has sucked, a lotion composed of one part strong tincture of *Hydrastis* and twenty parts water, being careful to wash it away before the child nurses again.

RHEUMATISM.

DEFINITION.—A specific, non-suppurative inflammation of the muscular, fibrous or serous tissues of the body, characterized by an acid state of the secretions and an excess of fibrin in the blood.

The following are the principal varieties :

ACUTE ARTICULAR RHEUMATISM.
(RHEUMATIC FEVER.)

SYMPTOMS.—Chills ; irregular fever, accompanied by sour sweating ; pain, soreness, stiffness and swelling of one or more of the joints. The tongue is coated, but the head and mind are but slightly affected. The inflammation often quits one joint and goes to another, or it may attack the heart, lungs, eyes or brain. Valvular disease of the heart generally originates in an attack of Acute Articular Rheumatism. Endocarditis occurs in about one-fifth of the cases and Pericarditis in about one-seventh of the cases of this form of Rheumatism.

CAUSES.—Peculiar predisposition ; exposure to cold and dampness. The poison of Rheumatism originates within the system.

TREATMENT.—The perspiration should be promoted by hot vapor baths, or blanket baths. The diet should be light and plain, consisting mostly of milk and farinaceous food.

Acon., in the early stage of the fever, may cut short the attack.

Rhus is a valuable remedy in the later stages, and in subacute forms of the disease.

Colchicum, in doses of five drops of the strong tincture, every four hours, is the most useful remedy known. *Cactus grandiflorus*, first dilution, should be given if there is pain in the region of the heart, or if there is palpitation.

CHRONIC ARTICULAR RHEUMATISM.

SYMPTOMS.—The joint pain may be constant for months or years, or, as is oftener the case, intermittent, coming on at every unfavorable change in the weather, or at every exposure to cold and dampness. There is fever, indicated by the frequency of the pulse, and constant perspiration ; the urine deposits a sediment. The joints may or may not swell, but are sensitive to pressure.

CAUSES.—Peculiar predisposition ; previous attacks of Acute Articular Rheumatism.

TREATMENT.—Keep the affected parts, and generally the whole body, clothed in flannel.

Rhus is one of the most valuable remedies, especially in the intermittent form of the disease.

Merc. iod. rub., taken internally three times a day, and used locally in the form of ointment once a day, will reduce the most obstinate swelling and at the same time moderate the soreness and pain. The strength of the ointment should be about one part of the salt to thirty parts of lard or simple cerate.

Kali iodatum may be used with good effect, after the last remedy has been tried for a week or two.

The digestion is generally at fault in this disease and should receive the greatest care. (See Dyspepsia.)

DEFORMING ARTICULAR RHEUMATISM.

SYMPTOMS.—At the beginning of the disease the pain is slight, but worse at night. The joints are enlarged and tender to pressure. A slight crackling is generally felt if the hand be laid on the joint when the limb is moved. The disease progresses slowly from bad to worse till the victim reaches a helpless old age. The patient may be confined to a chair or bed for years.

CAUSE.—The disease prevails mostly among the poor ; it attacks women more frequently than men ; and generally begins between the ages of thirty and forty years.

TREATMENT.—These cases are sometimes improved, but rarely if ever cured.

Merc. iod. rub., and *Kali iodatum*, used as recommended in Chronic Articular Rheumatism, will probably do more good than any other medicines. Warm bathing will be found very beneficial. The Hot Springs of Arkansas furnishes a promising resort to those who can afford the expense of a residence in that locality.

SALT RHEUM.
(ECZEMA RUBRUM.)

DEFINITION.—A variety of Eczema, or inflammation of the skin, characterized by superficial redness and the presence of numerous very small vesicles, which run together, burst and discharge a watery fluid, which dries into thin, yellow crusts.

SYMPTOMS.—Smarting or itching of the affected portion of the skin. The vesicles may be so small as not to be visible to the unaided eye. The skin is thickened and red. The discharge stiffens linen like starch. The most common sites of the disease are the scalp (Milk Crust) ; behind the ears ; the face ; the forearm : the back of the hand ; the legs.

The sudden suppression of the eruption may be followed by diarrhœa, Asthma, inflammation of the brain, or some other internal disorder.

TREATMENT.—The best internal remedies and their indications, are as follows :

Rhus,—Acute cases, with considerable redness and burning itching, and a watery discharge.

Sulphur,—Chronic cases, with very slight discharge and much itching.

Arsen.,—Chronic cases with burning itching,and a thin watery discharge, with the formation of thin crusts.

Croton tiglium, is one of the best indicated remedies, for this disease, in our *Materia Medica.* The third decimal dilution may be used internally, and a lotion of the second decimal, in water, in the proportion of one to ten, locally. The affected parts should be kept clean by frequent, gentle washing, with soft water. If soft water cannot be obtained, boiled water, to which has been added a little wheat bran, forms a good substitute. General bathing is of the greatest utility. A light application of *Cosmoline* is very useful in keeping the crusts soft and in preventing itching.

SCARLET FEVER.
(SCARLATINA.)

DEFINITION.—A specific, contagious, eruptive disease, characterized by continued high fever and the appearance of a fine scarlet rash, which begins in the mouth and throat and gradually extends over the skin of the whole body.

SYMPTOMS.—The primary fever is high. During the first day, the patient complains of headache, soreness of the throat and a pain in the stomach, and in some cases vomits once or twice. The eruption appears during the second day, in the mouth and throat, as a bright red rash which shows through the white coating on the tongue. During the third day, it appears on the neck, the upper part of the chest, the face and arms. During the fourth day, it covers the trunk, especially in front. During the fifth day, the legs and feet become covered. The duration of each crop of the eruption is about three days.

The fever does not subside immediately on the appearance of the eruption, as it does in Measles, and in Small-pox, but even rises during the coming out of the rash and after the eruption is complete. At the hight of the fever the temperature of the blood is from 103° F. to 105° F., or in fatal cases, 109° F. to 112° F. The pulse beats at the rate of 120 to 150 per minute.

In about seven days from the beginning of the attack the outer layer of the skin begins to peel off. This process of desquamation lasts from eight to fourteen days, making the total duration of the disease about three weeks.

Although the poison of Scarlet Fever is always the same, authors make four varieties of the disease, according to its severity.

1. *Irregular* or *imperfect* cases, in which either the eruption or the fever or the soreness of the throat is wanting. Such cases cannot be recognized as those of Scarlet Fever except when it is known that the patient has been exposed to the disease.

2. *Simple Scarlet Fever—Scarlatina simplex,*—in which although there is some soreness of the throat, there is no difficulty in swallowing, or swelling of the glands.

3. *Anginose Scarlet Fever—Scarlatina anginosa,*—in which the throat symptoms are well marked and the cervical glands are swollen.

4. *Malignant Scarlet Fever—Scarlatina maligna,*—in which the inside of the throat is covered with a whitish exudation resembling that of Diphtheria. The nose and even the eyes may be affected in the same way as the throat. Such cases are almost always fatal.

The mortality of this disease varies greatly in different epidemics. In some seasons and places only one in thirty die ; in others, as many as one in four.

CAUSE.—Nearly all cases of Scarlet Fever can be proven to have been produced by direct or indirect exposure to the emanations from the bodies of those suffering from the disease. The peculiar poison, like that of Small-pox, may be conveyed in articles of clothing or retained in bedding, carpets, upholstery, or wall paper, for an indefinite length of time. The first symptoms appear in about seven days from the exposure. Children from the age of two to five years are most susceptible to the disease. After the age of five years the susceptibility declines, but the disease may attack persons of any age. It is exceedingly rare for the same person to be attacked twice. Many persons go through life with frequent exposures and never contract the disease.

PREVENTION.—No person should be unnecessarily exposed to this capriciously fatal disease. Rooms and houses in which it has existed may be disinfected by exposure to the vapors of burning Sulphur. Small vessels of Sulphur may be ignited in the different apartments and left to burn, while the doors and windows are closed and the occupants outside. After several hours the doors and windows may be opened and the house thoroughly ventilated. Articles of silver or plated with silver, should be removed from the house while this process is going on, lest they be blackened. Clothing may be disinfected by boiling, baking or burning. The vaunted preventive powers of *Bella-donna, Sulpho-Carbolate of Sodium, Camphor*, and other medicinal drugs, have not been verified by the rigid test of experience.

TREATMENT.—In the milder cases, *Acon.* and *Bell.* are the only medicines required. The latter may be given till the stage of desquamation begins.

In the anginose form, *Merc. iod. rub.* is an exceedingly valuable medicine.

In the malignant form, when there is prostration of the vital powers, *Chininum arsenicosum*, three grains of the third decimal trituration, in alternation with *Merc. iod. rub.*, will probably do more than any other remedy.

Ailanthus glandulosus has lately been highly recommended on theoretical and clinical grounds. It is peculiarly adapted to the anginose and malignant forms of the disease, especially if the eruption is dark, the throat inclined to be gangrenous, and if there is a watery discharge from the nose. The dropsical symptoms which sometimes appear in the third and fourth week, are best relieved by the use of *Apis mellifica*, third, and *Arsen.*

The room should be well ventilated. To attain this end there must be a constant current of pure air flowing through the room. This may be done in such a manner as not to chill the patient. The bed-clothing should be frequently changed. Plenty of cold water and milk should be allowed for drink and nourishment. Frequent tepid baths, without friction in drying, will comfort the patient, diminish the fever, and improve the chances for recovery. There is no danger of giving a chill, if the exposure is not *prolonged*. The amount of cold which can be borne with safety depends upon the hight of the fever and the strength of the patient.

SCROFULA.

DEFINITION.—A constitutional disease, characterized by defective nutrition and softness of the tissues.

In the acute forms of the disease, albuminous matter, generally of a cheesy consistence, is deposited in various parts of the body. These deposits are called tubercles. When they take place in the lungs, *Consumption* is the usual result. They may, however, form in the brain or its membranes, producing *Dropsy of the Brain*; in the pleura, in the throat, the bronchial tubes, the liver, the spleen, the peritoneum, or the mucous membrane of the bowels.

In the chronic forms of the disease, there is swelling of the lymphatic glands in the neck and in the groins, chronic inflammation of the eyes, hip joint disease, White Swellings of the joints, decay of the bones, chronic discharge from the ears.

SYMPTOMS.—Loss of appetite, especially for fats; irregular appetite; persistent fever; (the temperature ranges from the normal $98\frac{3}{4}°$ F. to 101° F., or in rapidly progressive cases, 103° F.); slow and gradual loss of flesh; cold perspiration of the hands and feet; night sweats. The enlarged glands suppurate, discharging thin pus, and heal slowly.

CAUSES.—Scrofulous or sickly parentage; breathing bad air; enervating, or unwholesome habits of life.

TREATMENT.—Children predisposed to scrofulous disease should breath only pure, fresh air, night and day. A residence near the seaside and the practice of sea-bathing will assist in preventing the development of the disease. The patient should be put upon full diet of the most nutritious and digestible foods. Beef, milk, poultry, eggs, bread and butter with a small allowance of mealy potatoes should form the greater portion of the food.

The following remedies will be found useful as they are indicated:

Arsen.—Indigestion, with sour stomach or bloating, cold sweat of the hands and feet, night sweats.

Nux.—Indigestion with constipation or irregularity of the bowels.

Phos.—Short, dry cough ; choking sensation in the upper part of the chest ; nervous debility.

Merc. iod. rub.—Enlarged glands.

Silicea, sixth decimal trit.—Suppurating glands, which are slow to heal.

Puls.—Indigestion, with pain in the stomach and nervous symptoms.

SMALL-POX.
(VARIOLA.)

DEFINITION.—A specific, contagious, eruptive disease, characterized by a high fever, followed by the development of pustules on the skin, each of which, when mature, has a depression in the center, and after healing leaves a white scar.

SYMPTOMS.—The first symptoms of the disease appear twelve or thirteen days after exposure. The preliminary fever comes on with a chill and is accompanied by intense headache, especially in the back of the head, vomiting, pain in the back and an inclination to drowsiness. The eruption appears on the fourth day of the fever. It appears, first, in minute, bright-red specks on the face, neck, upper extremities, trunk and lower extremities. These soon fill with a watery fluid, enlarge, and, on the eighth day of the disease, begin to "maturate." The skin between the pustules then begins to be inflamed, and the fever, which had subsided on the appearance of the eruption, again rises and continues high till the twelfth day, when the pustules burst and form crusts which fall off about the fourteenth day of the disease. The number of pustules varies from twenty to many thousands ; the fewer the pustules the lighter the disease. In the "confluent form," the pustules are so near together as to mingle when they fill and produce one immense covering of corruption. These cases are almost always fatal.

CAUSE—Personal contact with the disease. The infecting distance varies with the amount of poison in the air and the direction of the wind, from five to one hundred feet, or perhaps it has even wider limits. The disease is most contagious when fully developed. The clothing, as well as the crusts, retain their infecting properties for many months, especially if kept in a cold place.

One attack of the disease gives immunity from subsequent attacks. Extensive observers conclude that only one in one hundred of those who have had the disease will contract it again on exposure.

PREVENTION.—The means of disinfection recommended in the article on Scarlet Fever are efficacious against the Small-pox poison.

Vaccination changes the character of the disease which may supervene on exposure to Small-pox. The following table, showing the results obtained by Mr. Simon, from a collation of 6,000 cases, shows the percentage of deaths in cases of Small-pox, in each of the six classes given:

I.	Unvaccinated......	33⅓	per centum.
II.	Unsuccessfully vaccinated..................	21¾	" "
III.	Having one vaccine cicatrix..................	7½	" "
IV.	Having two vaccine cicatrices	4½	" "
V.	Having three vaccine cicatrices......	1¾	" "
VI.	Having four or more vaccine cicatrices......	¾	" "

Vaccine lymph should, if possible, be taken directly from the animal. It may be inserted in the arm, about midway between the shoulder and the elbow, in three or four places, about an inch apart. The more virus there is absorbed, the more certain and effectual will be the vaccination. One of the best methods of vaccinating is to make four or five parallel cuts or scratches, each about one-fourth of an inch in length, with the point of a needle or a sharp knife, just deep enough to cause a slight flow of blood; then apply the virus (which, if dried, should have been just previously moistened with cold water) by rubbing it transversely to and fro across the cuts. On the eighth day after the operation the vesicles should be filled with lymph, circular in shape, with the rim higher than the center.

TREATMENT.—In favorable cases only the three following remedies will be required. In the first stage, *Acon.;* in the eruptive stage, *Tartar emet.* and in the drying-up stage, *Sulphur.*

If there is active delirium, give *Bell.* If the eruption should suddenly disappear, give three drops of the strong tincture of *Camphora.*

To prevent pitting, smear the face and neck with a paste made of flour and sweet cream.

The importance of *fresh air* cannot be to much insisted on. In warm weather, there should be a constant current of air through the room, and, in cold weather, a window should be left up and the. patient protected by a light, warm covering. The posture in bed should be frequently changed to prevent bed-sores. The sheets should be kept clean and smooth.

SNAKEBITES.

The deadly poisonous serpents of the United States are: the *Rattlesnake* (*Crotalus durissus*), and the *Copperhead* (*Trigonocephalus contortrix*). Both have a triangular shaped head and a smaller neck, and two poison-fangs in the upper jaw. The former may be known by the noise of its rattles, and the latter by its copper colored eyes.

TREATMENT.—A handkerchief or cord should be tied tightly around the limb between the wound and the heart.

It is well to suck the wound, so as to extract the poison, provided there are no abrasions in the mucous membrane of the mouth.

Whisky and other alcoholic drinks, tend to counteract the effect of the poison until it can be eliminated from the blood. Much larger quantities are required to produce alcoholic intoxication than if the bite had not been received. If it can be done within an hour from the time the bite was received, the wound should be seared with a hot iron or some caustic. The quicker this is done, the more likely it is to be of service.

SPERMATORRHŒA.

DEFINITION.—A diseased condition in which there are frequent discharges of seminal fluid, unprovoked by voluntary excitation.

SYMPTOMS.—The discharges occur generally during sleep, but may take place while the patient is straining at stool, or at various other times. They vary in frequency from once in two or three weeks, to several times in the course of twenty-four hours. Cases in which the general health is vigorous, and the discharges do not occur oftener than once in ten days, are not subjects for medical treatment.

The effects of this disease are glowingly portrayed by advertising charlatans, who often gain a handsome income from the profits of this business, while their deluded victims are seldom benefited.

The most common results of the disease are: depression of spirits; forgetfulness; lack of bodily vigor; indigestion; constipation; and, in rare instances, sexual impotence and mental imbecility.

CAUSE.—Self-abuse is by far the most frequent cause. The practice is usually begun at, or soon after, puberty, when the boy is ignorant of its consequences. Constipation, injuries to the spine, and the presence of worms in the rectum have been accused of producing the disease.

PREVENTION.—Proper and timely instruction in the physiology and pathology of the functions of the generative organs, would in most cases be sufficient to prevent youths from contracting this degrading habit. Idleness, solitary habits and debasing companionship should not be allowed during the age in which the manly functions are developing.

TREATMENT.—If the disease has become established, all the habits of living must be brought to the strictest regularity. Stimulants and late suppers must be especially avoided. The diet however, should be rich and sustaining. The patient should accustom himself to sleep on the side, and care should be taken that the bed-clothing be not too warm. Constipation should be carefully guarded against. A good rule is to have a passage from the bowels after the last meal of the day.

Nux is a very useful remedy in the treatment of this disease.

Three drops of tincture of *Gels.*, taken on retiring, will many times prevent an emission. The application of *Belladonna ointment* to the perineum will effect the same and with less constitutional depression. Extreme cases require careful medicinal and sometimes surgical treatment, but this should never be sought except of a regular physician.

SPRAIN.

DEFINITION.—An injury to the ligaments or tendons, produced by overstretching. A similar injury to the muscles is called a *Strain*. There may or may not be an appreciable rupture of the fibres or attachments.

TREATMENT.—Apply cloths wrung out of *hot water*, till the acute pain is relieved. Then bandage the joint and moisten the bandage with *Arnica lotion*, made by adding ten teaspoonfuls of water to one teaspoonful of *Arnica* tincture. When the soreness is so much reduced that slight movements do not cause pain, the dressings may be removed and replaced by strips of *Surgeons' Adhesive Plaster*, so applied as to prevent any movement of the joint which causes pain. Perfect rest for a week or more is essential to the complete and rapid cure of a Sprain.

A *muscular Strain* may be better treated by using *Arnica* internally. The dose is one drop of the tincture or dilution. The tincture of the root is preferable.

STINGS OF INSECTS.

SYMPTOMS.—The effects of the stings of bees, wasps, mosquitoes and other insects, vary much in severity in different persons. In some individuals the sting of a bee produces only the slightest local irritation, while in others the most alarming constitutional symptoms may arise. The following remarkable case, which occurred in the practice of Dr. R. C. Sabin, of Milwaukee, Wis., illustrates the extreme susceptibility of some persons to the poison of the bee:

"A young lady, aged about seventeen, was stung by a honey bee on the forehead, over the left eye, while playing croquet. She screamed with pain and tried to tell the bystanders what had happened to her. Her speech was very indistinct. I was in the house at the time, and on reaching the patient found her lying on the grass. I observed that the tongue was swollen to such an extent as to project from the mouth. The lips and gums were also swollen, and of a deep blue color. There could not have been more than three minutes' interval between the receipt of the sting and my examination of the case. The poor girl was alarmed and almost wild with pain; could not control herself; seemed hysterical and had to be carried to her room. She complained of excruciating, cramp-like pain in the stomach, which caused her to bend double and press with her hands upon the epigastric region.

I gave her a tumblerful of a mixture containing equal parts of whisky and hot water, which was swallowed with difficulty, but produced prompt relief from the pain. In a few minutes she fell into a deep sleep. This was about twenty minutes from the time she was stung. She slept for about two hours, and when she awoke the swelling of the tongue, gums. etc., had entirely disappeared; but the limbs below the knees, the neck and upper part of the arms were thickly covered by an eruption closely resembling that of Measles. She informed me that the eruption remained over twelve hours, fading out almost imperceptibly. There was no inflammation at the seat of the injury, until the following day, when it began to show itself, and eventually reached a high degree of severity, the eye being completely closed for three days. She had no constitutional symptoms after the first twelve hours.

The effects in this case were the severest I ever witnessed, but I am acquainted with two other persons in whom the sting of a bee is always followed by pain in the stomach, vomiting, swelling of the tongue and general anasarca."

TREATMENT.—The application of *very hot water*, by means of a cloth folded to a point, will coagulate the poison and check the inflammatory action. The local application of *Alcohol* will effect the same result. Although the stings of insects as well as the poison of serpents are slightly acid, their poisonous properties are not destroyed by alkalies.

STYE.

DEFINITION.—A small Boil on the eyelid.

SYMPTOMS.—Soreness, itching, pain, heat, redness and swelling on the lid, sometimes accompanied by fever and headache. When pus forms it is usually discharged at the edge of the lids.

CAUSES.—A depraved condition of the blood. It most frequently occurs in young persons of scrofulous tendency and impaired digestion.

TREATMENT.—*Acon.* and *Puls.* in the acute forms, and *Rhus* and *Sulphur* in the chronic forms, will be found most generally curative. If the pain is severe, the application of warm water dressing, or warm poultices, will give relief and hasten suppuration.

SUNSTROKE.

DEFINITION.—Paralysis of the brain, produced by excessive heat.

SYMPTOMS.—These may come on slowly or suddenly. There is languor, weakness, dizziness and tendency to stupor. The face is flushed and the eyes are congested. In this state a sudden attack of *Syncope* may cause death any moment. The temperature of the brain is several degrees above the normal standard of 98° to 99° F. Congestion of the lungs occurs in almost every case.

TREATMENT.—Remove the patient to a cool place; strip the clothing from the head, neck and upper part of the chest, and apply cold to the neck and head. This may be done by pouring on cold water, or by applying cloths wrung out of cold water, or bags of pounded ice.

If the pulse is feeble and the face pale, give a dose of *Camphora,* or hold a bottle of *Spirits of Camphor* to the nose.

If the face is red and the pulse full, give *Bell.*

SYNCOPE.
(FAINTING.)

DEFINITION.—A sudden loss of consciousness and muscular power, due to functional derangement of the brain.

CAUSES —The immediate causes of fainting are strong mental emotions, loss of blood, disease of the heart, general debility, indigestion, Hysteria, etc.

TREATMENT.—Lay the body in a horizontal position, with the head low; sprinkle or dash cold water on the face; loosen the clothing. If the patient should not revive after a few seconds, apply a bottle of *Camphor* or *Ammonia* to the nostrils.

SYPHILIS.

DEFINITION.—A chronic disease, resulting from a specific poison which is communicable, either by inheritance, or by actual contact with a mucous or an abraded cutaneous surface.

SYMPTOMS.—The disease first appears as a papule, pimple, fissure, abrasion or crack, around which there arises in a few days a hardened ring. The surface enclosed by this ring generally ulcerates and becomes covered by an exudation of an ashy-gray color. This sore, which is technically called a "Chancre," is about one-fourth of an inch in diameter. It appears from the third to the tenth, usually in the fourth, week after the deposit of the poison. The most common locations of chancre are on the mucous membrane of the genital organs, on the lips, gums, tongue or eyelids. In the course of four to twelve, usually six, weeks after the first appearance of the primary sore, the glands of the groins and of the neck enlarge and become hard, but not very sore, and red spots appear on the face and breast. In about six weeks more, there develops a peculiar chronic or subacute sore throat, and an eruption of raised, dark spots on the face and various other parts. A scurfy eruption covers the hairy scalp.

After the subsidence of these and a few other symptoms, classed as secondary, there is a comparative lull in the course of the disease for some months or years, when there begins a more irregular and more alarming train of effects. Tumors grow upon the bones; the mouth and throat are the seat of a series of ulcers which may even destroy the palate, the jaw bones and parts of the throat; specific deposits form in the lungs, liver and internal organs. The disease may destroy the life of the patient in one of many ways, such as the growth of bony tumors pressing on the brain or spinal cord; the deposits in the lungs, producing a form of Consumption; and the gradual wearing out of the vital energies, by the formation of extensive ulcers. These and many other symptoms constitute what is called the tertiary stage. In the inherited form, the symptoms run a course somewhat different. The primary stage is absent; the secondary begins at the age of two to four weeks, and lasts several months. Diseases of the skin are the most prominent feature. The tertiary stage may begin at the time of the development of the permanent teeth, or at any later period. The eyes suffer from a peculiar form of inflammation. Nervous deafness is quite common. The teeth have a peculiar notched appearance.

PREVENTION.—Avoid personal contact, especially of the mucous surfaces, with persons known or suspected to have the disease. Even the use of spoons and drinking vessels has been the means of communicating the disease. If by any means the poison of *Syphilis* has been deposited on any part of the body, wash the part immediately and thoroughly. If a primary sore be discovered before the hardened ring has formed, or the glands have enlarged, cauterize the sore by the application of strong *Nitric Acid.* This may destroy the poison before it has arrived at such a stage of growth that it can develop in the blood and produce the constitutional symptoms.

TREATMENT.—A respectable and regularly educated physician and surgeon only, is competent to conduct the treatment of this disease. The common practice of consulting advertising quacks in reference to the so-called "Venereal diseases" is extremely hazardous as well as absurd. These ignorant bunglers add to the horrors of this specific blood-disease the baneful effects of drug-poisoning, and hasten to an untimely grave many more than they cure.

Merc. viv.,—second decimal trituration, in two to five grain doses four times a day, will, in the earliest stages, do more than any other remedy toward the elimination of the poison, and the prevention of its constitutional effects.

Merc. iod. rub.—will be found more serviceable in the secondary stage, after the induration and glandular enlargments have taken place. In the tertiary stage, this is a remedy against the bony tumors.

Kali iodatum—may be advantageously alternated with *Merc. iod. rub.*, if the latter requires to be given for a long period of time. This remedy assists in the elimination of the mercurial preparation, as well as reduces the indurations.

Nitri Acidum,—second decimal dilution, is useful when the chancre is slow to heal, but is especially demanded in the treatment of the ulcerations of the tertiary stage.

Arsen. may be required if there are sloughing gangrenous ulcers and an impaired state of the general health.

Merc. cor. may be given in case there is a urethral or vaginal discharge, or if there are corroding ulcers with a watery secretion.

Scrupulous cleanliness will do much in the way of removing the irritant poison as well as in preventing its access to the circulation.

TEETHING.

The milk teeth, or temporary teeth, are twenty in number. They break through the gums in sets of four, the two lower preceeding the two upper, in the following order.

```
7th month,—Central incisors, or first front teeth.
7th to 10th month,—Lateral incisors, or second front teeth.
12th to 14th month,—Anterior molars, or first grinders.
14th to 20th month,—Canine, or eye-teeth.
18th to 36th month,—Posterior molars, or last grinders.
```

There may be some variation from this order and time without ill health.

The symptoms and treatment of some of the most common disorders caused by teething are as follows:

Soreness of the Gums.—Apply with the finger a lotion of *Arnica* of the strength of one part of the strong tincture to ten parts of water. Lancing the gums is very seldom necessary.

Feverishness.—Acon.

Fretfulness.—Chamomilla.

Sleeplessness.—Gels., in one-fourth drop doses.

Diarrhœa.—Merc. viv., or if this does not soon relieve, *Hydrastis*, in drop doses, of the strongest tincture.

Bronchitis.—Tart. emet.

Infants should not be weaned while there is irritation of the gums. The best time for weaning is after the appearance of the first eight teeth.

TOOTHACHE.

CAUSES.—Decayed teeth; exposure to cold; Neuralgia. Decay of the teeth generally proceeds from habitual indigestion, in which there is an acid state of the fluids of the mouth, or from taking too much acid into the stomach in the form of food and drink.

TREATMENT.—If the tooth is hollow, introduce some cotton-batting moistened with *Creosote*, or strong tincture of *Aconite*. If the tooth is not decayed, the same preparation, in a diluted form, may be rubbed on the gums around the root. If the pain is periodical, give *Arsen.* internally. If there is fever give *Acon.* or *Bell.*

ULCER.

DEFINITION.—A surface deprived of its proper covering of skin or mucous membrane. The immediate
effects of a recent bruise or burn are not usually called
ulcers, but in depraved constitutions they may be soon
converted into ulcers.

CAUSES.—Bruises ; burns ; prolonged pressure ; obstructed circulation through the veins ; diminished
power of nutrition ; contagious poisons, as in Chancroid,
and some forms of sore mouth ; Scrofula ; Syphilis.

TREATMENT.—*Arsen.* may be given if the ulcer has a
livid appearance and gives off a thin discharge.

Silicea is indicated in case the ulcer is chronic and
torpid, and the discharge thick and yellow.

Nitri acidum, is the remedy if the sore shows a
tendency to eat or spread, especially if there is a syphilitic taint, or if the patient has taken large quantities of
Mercury.

Kali iodatum, is more appropriate if there is a scrofulous taint.

The local treatment is very important. Scrupulous
cleanliness should be kept up by frequent washings with
water. The sore should not be cleansed by rubbing, but
by letting a stream of water fall on it. Frequent
ablutions of the feet and legs will help to prevent the
development of ulcers.

Firm, continuous pressure will greatly promote the
healing of ulcers, both by bringing the edges near
together and by supporting the blood-vessels.

The most convenient means of applying pressure is
the application of strips of *Surgeons' Adhesive Plaster*. If

a limb is to be bandaged, the strips should be no more than an inch in width, and the strapping should be begun at the farthest extremity and be carried up to the ulcer. It is well to lay a piece of tin-foil immediately over the raw surface. If there is a fœtid discharge, a dressing of *Carbolated Cosmoline* or a lotion of *Carbolic Acid*, of the strength of one to one hundred, may be applied whenever the dressings are removed. The plasters ought to be changed every two or three days.

URINARY DISORDERS.

RETENTION OF THE URINE.

DEFINITION.—Excessive accumulation of urine in the bladder.

CAUSES.—Obstruction to the passage of water ; paralysis of the bladder. Obstruction of the urethral passage may be caused by Spasmodic Stricture, Organic Stricture, or enlargment of the prostate gland.

SYMPTOMS.—Pain, tenderness and fullness in the region of the bladder ; dull sound, produced by tapping over the bladder ; diminished quantity of urine voided. Spasmodic stricture comes on suddenly after exposure to cold, or partaking of improper drinks. Organic stricture comes on gradually, some months or years after an attack of Gonorrhœa. Enlargment of the prostate gland is most common in persons over sixty years of age.

TREATMENT.—Spasmodic stricture may be relieved by the administration of one or more drops of *Camphor* or *Gels.*, and the application of cold water to the perineum. Organic stricture requires the introduction of a catheter into the bladder. The instrument should be as large as can be introduced, and should be well oiled and warmed before use. The cure of organic stricture is accomplished by the regular and persistent introduction of larger and larger conical bougies. The instrument ought to be introduced about once in two days, and its use continued for some time after the easy introduction of a "No. 12."

SUPPRESSION OF THE URINE.

DEFINITION.—Diminished secretion of urine by the kidneys.

CAUSES.—Disease of the kidneys; febrile conditions.

SYMPTOMS.—Diminished quantity of urine voided; pain in the back and loins; fever; hollow sound, produced by tapping over the region of the bladder.

TREATMENT.—*Acon.* is the best remedy for the general condition of fever, which causes diminished secretion.

Apis mellifica or *Arsen.*, will be found useful, if there is dropsy.

INCONTINENCE OF URINE.

DEFINITION.—Involuntary or unconscious discharge of urine.

CAUSES.—Irritability of the neck of the bladder; inflammation of the bladder; paralysis of the bladder; worms in the alimentary canal.

SYMPTOMS.—The trouble may be constant, indicating Chronic Inflammation of the Bladder; or acute, indicating Acute Inflammation; or mostly at night, indicating simply irritability of the neck of the bladder.

TREATMENT.—*Gels.* will be found curative if there is general irritability or fever.

Canth. is preferable, if there is some pain in passing water.

Equisetum hyemale acts like a charm in cases of children who are in the habit of wetting the bed at night. The dose is two to four drops of the strong tincture on retiring.

PAINFUL URINATION.

CAUSES.—Inflammation of the urethra (Gonorrhœa); Gravel.

TREATMENT.—The following are the remedies :

Acon.,—trouble caused by taking cold.

Camphora,—severe acute cases.

Canth., —burning or smarting in the urethra.

Merc. cor.,—discharge of mucus with the urine.

PROFUSE URINATION.

CAUSES.—Hysteria ; Diabetes.

DIAGNOSIS.—In Hysteria, the specific gravity of the urine is below 1.020.   In Diabetes, the specific gravity is above 1.030.   The normal quantity of urine for an adult is from two to three pints, in the course of twenty-four hours.

VOMITING.

(SEE NAUSEA.)

Simple vomiting of food without nausea or retching, requires no treatment.   It is produced by over filling the stomach.

A sudden and forcible projection of watery fluid from the stomach, not preceded by nausea or accompanied by retching is a symptom of irritation of the brain. *Bell.* is the proper remedy.   In case this fails give *Helleborus niger.*, third decimal.

WARTS.

When a number of small warts appear at nearly the same time, it is proper to assume that there is a peculiar condition of the system which favors the growth of these excrescences. *Thuja occidentalis*, first dilution, taken internally, for a week or two, will probably suffice for their removal. Solitary, old, hard warts are difficult to remove. The local use of lye from wood ashes will soften the wart, and if continued persistently, and aided by occasional pricking and scraping, will remove the wart. *Sulphur* is a good remedy to prevent the recurrence of the growth.

WHOOPING-COUGH.

DEFINITION.—A specific, contagious disease, characterized by a peculiar cough, which consists of paroxysms of short, forcible expirations, followed by one long, deep inspiration.

SYMPTOMS.—The first symptoms resemble the effects of a cold. The peculiar spasmodic cough develops after the catarrhal symptoms have existed about a week. A paroxysm begins with slight tickling in the throat, which excites a cough. The patient, by several short, successive expiratory efforts, gradually expels the air from the lungs, and continues the process till the face is dark red, the eyes are suffused and suffocation seems imminent ; then, by one long inspiratory effort, generally accompanied by a loud, harsh " whoop," the lungs are filled and the paroxysm is over.

A small quantity of tenacious mucus is detached from the bronchial tubes and expectorated by this pro-

11

cess. The severity of the disease may be estimated by the frequency of the paroxysms. If they occur only four or five times in the course of twenty-four hours, the case is a mild one ; but if they come as often as once an hour, the disease is severe. Children take the disease more readily than adults, and it generally affects them more seriously.

TREATMENT.—The diet should be light and of easily digested food. The stomach should never be filled to repletion, either by food or drink. The patient should take food every two or three hours, rather than overload the stomach. *Gum Arabic, Slippery Elm bark,* or *flax-seed,* may be put into the drinking water, to promote an easy expulsion of the mucus which clings to the bronchial tubes.

Ipecac and *Nux* are the remedies which are most frequently curative. *Kali bichrom.* is a valuable aid to the clearing out of the air passages. *Kali bromatum,* in doses of one-tenth of a grain to one grain, every hour, according to the age and strength of the patient and the severity of the cough, will greatly moderate the paroxysms. The worst paroxysms may be broken up by the inhalation of a drop or two of *Amyl nitrosum,* placed on the palm of the hand.

The worst cases may require a change of residence to a mountainous or a seaside locality.

WORMS.

Of the thirty-one species of worms which are known
to infest the human body, only four are found so frequently as to demand a notice in this work.   These are :

1. *Tænia solium*, (*Linnæus*), —Tape Worm.

2. *Ascaris lumbricoides*, (*Linnæus*),—Long Round
Worm.

3. *Oxyuris vermicularis*, (*Bremser*),—Pin Worm.

4. *Trichina spiralis*, (*Owen*),—Pork Worm.

TAPE WORMS.

DESCRIPTION.—The *fully developed Tape Worm* is composed of 800 to 1,000 segments, and measures from 9 to
35, or, in some cases, more than 100 feet in length.
At the end farthest from the head, it is from one-third
to one-half of an inch in width.   Toward the head it
tapers to a mere line, and the head itself is so small as
to be seldom seen.

The *head* is furnished with four suckers or mouths,
and has at its very tip a double row of minute hooks,
about thirty in number, by which the animal retains its
hold on the intestine.

The *segments*, when fully developed, are about three-
fourths of an inch in length and one-third of an inch
in breadth.   Each one contains the generative organs of
both sexes, and after the fecundation and growth of its
eggs, becomes detached from the worm and is carried
out during the movements from the bowels, or at other
times.   Occasionally several are detached at once.   The
eggs remain in the segment till the latter is destroyed by

the action of heat and moisture. They may then be blown about by the winds, or float in water for a long time without losing their vitality. Occasionally they are taken into the stomach of some animal, along with its food. The leathery husk or shell is then dissolved, and the young animal begins to develop as a minute, slender worm, which penetrates the walls of the stomach and finds its way into some of the solid tissues, such as the liver or the mesentery.

Here it becomes enclosed in a sac and forms what is called a "*Hydatid Cyst.*" Within this cyst it grows to some length and the segments of a future Tape Worm become marked out by lines across the body. The disease called "Staggers," which occurs in sheep and cattle, is caused by the growth of these Hydatid Cysts in the brain. When these Hydatid Cysts are eaten by man or other animals the worm is set free, and by its little hooks, which have already grown on the head, it fixes itself upon the mucous membrane of the intestines, where it begins a new round of existence.

SYMPTOMS.—Spasmodic, gnawing, biting pain in the abdomen; itching of the nose, mouth and anus; very dark, or very light stools; capricious appetite, sometimes absent, sometimes voracious; disturbed sleep; irritable disposition.

PREVENTION.—Strict abstinence from the eating of *raw meat* is the only sure preventive of Tape Worm. *Raw vegetables* grown where dogs have free access, should be avoided, for fear of the introduction of Tape Worm eggs and the development of Hydatid Cysts. For the

same reason, all Tape Worms and segments, not preserved for scientific purposes, should be thoroughly
destroyed by fire.   Dogs should never be allowed to eat
the flesh of animals which have died of Staggers, lest
the dogs become infested with Tape Worms, and the eggs
of the latter fall into the food and drink of the sheep
and produce in them the Hydatid Cysts.

TREATMENT.—Two or three teaspoonfuls of tincture of
Rottlera tinctoria, taken after twelve hours' fasting will
rarely fail to kill and expel the worms.   A dose of *Castor
Oil* may be given if no purgative action follows the
taking of the medicine.   This remedy is the pleasantest,
safest and surest with which I am acquainted.   The
tincture must be of good quality.

LONG ROUND WORMS.

DESCRIPTION.—The full-grown worm is from six to
sixteen inches in length, pointed at both ends and of
a grayish red color.   This species inhabits mostly the
intestinal canal, but individuals are sometimes found
strayed into the gall ducts, or into the stomach, œsoph-
agus, mouth, nostrils or even into the frontal sinus.   The
fertility of this worm is astonishing.   Dr. Eschricht
has made an elaborate calculation, by which he esti-
mates the number of eggs in a single mature female to
be 64,000,000.

It is highly probable that the eggs of this worm may
be introduced into the stomach by means of drinking
water which contains them.

SYMPTOMS.—The symptoms are very similar to those
produced by the Tape Worm.   The round worms are how-

ever, more common in children than in adults, and sometimes produce fever, and even dangerous irritation of the brain. A short, dry cough and a white ring about the mouth are common worm symptoms. The only unmistakable sign is the appearance of the worm itself.

TREATMENT.—*Santonin*, second decimal trit., in doses of one to three grains, before meals and at bed-time, is generally an efficient remedy. If worms are known to be present and this fails, the first trit. may be used in the same manner. *Merc. viv.* is said to be useful as a preventive.

PIN WORMS.

DESCRIPTION.—These are small, white, thread-like worms, the male being about one-eighth of an inch, and the female about a half an inch in length. Their common habitat is the rectum and the lower part of the small intestine, but they sometimes migrate to the vagina, or urethra of the female, or even to the rectum of the patient's bedfellow. The eggs are deposited, and the young are hatched near the anus. The worms require a supply of external air.

SYMPTOMS.—Itching of the anus, especially at night; incessant desire to go to stool.

TREATMENT.—Inject a pint of transparent *Lime-Water* into the bowels every night for a week, and every alternate night for two weeks more, and keep the anus and surrounding parts smeared with *Lard* or some other oily substance.

PORK WORMS.

DESCRIPTION.—The *mature worm* is very minute, the male measuring about $\frac{1}{15}$, and the female $\frac{1}{8}$ of an inch in length. Its home is in the intestines of warm-blooded animals, including man.

The young are born alive, and in a few days penetrate the walls of the stomach and intestines and enter the muscles of the abdomen, chest and shoulders. Here they grow for about fourteen days and then become enclosed in cysts or sacs, which are large enough to be seen by the unaided eye. The young Trichinæ in these sacs are about $\frac{1}{36}$ of an inch in length. They lie coiled up, one, two or three in each sac.

If the muscular tissue containing these sacs of Trichinæ, be eaten by man or animals, the capsules are digested and the worms set free. A brood of young are ready in about a week to set out for the muscles.

SYMPTOMS.—Nausea, distress in the stomach, diarrhœa, followed in a few days by great pain, soreness and stiffness of the affected muscles. The muscles of the chest, abdomen and shoulders are most affected. When the patient recovers, the stiffness of the muscles lasts for a long time and the general health is poor.

PREVENTION.—Eat no pork which has not been thoroughly cooked or long salted. Partially cooked sausage and ham are the most frequent means of producing the disease. Hogs generally contract the disease by eating rats ; the latter animals being very frequent sufferers from the ravages of Trichinæ.

Salaried pork inspectors cannot be relied on to exercise the constant vigilance necessary to prevent the sale of " measly pork."

TREATMENT.—If the introduction of Trichinæ into the stomach is discovered before the muscles are affected, give full cathartic doses of *Castor Oil* for three or four successive days. This will remove many of the young worms before they are old enough to migrate from the intestinal canal.

No poison has been discovered, which will kill Trichinæ in the muscles, without killing the patient.

Part II.

MATERIA MEDICA.

LIST OF REMEDIES FOR A MEDICINE CASE.

	Name.	Strength.	Form of Preparation.
1.	ACONITUM,	3	Dilution or pellets.
2.	Apis,	3	Trituration.
3.	ARNICA,	3	Dilution or pellets.
4.	ABSENICUM,	4	Trituration, dil. or pel.
5.	BELLADONNA,	3	Dilution or pellets.
6.	BRYONIA,	3	Dilution or pellets.
7.	Cactus,	2	Dilution or pellets.
8.	CAMPHORA,	0	Tincture, one in five.
9.	CHAMOMILLA,	3	Dilution or pellets.
10.	Chininum arsenicosum,	3	Dilution or trit.
11.	Coffea,	3	Dilution or pellets.
12.	Colchicum,	0	Tincture, one in three.
13.	Dulcamara,	3	Dilution or pellets.
14.	Erigeron,	0	Tincture, one in two.
15.	Ferrum phosphoricum,	1	Trituration.
16.	GELSEMIUM,	0	Tincture, one in three.
17.	HAMAMELIS,	0	Tincture, one in three.
18.	Hepar Sulphuris,	6	Trituration.
19.	Ignatia,	3	Dilution or pellets.
20.	IPECACUANHA,	3	Dilution or pellets.
21.	Iris,	3	Dilution or pellets.
22.	KALI BICHROMICUM,	3	Trituration.
23.	Kali bromatum,	0	Pure salt, pulverized.
24.	Kali iodatum,	2	Dilution or pellets.
25.	MACROTIN,	3	Trituration.
26.	MERCURIUS COR.,	3	Dilution, pel. or trit.
27.	MERCURIUS IOD. RUB.,	3	Dilution, pel. or trit,
28.	Mercurius vivus,	3	Trituration.
29.	NUX VOMICA,	3	Dilution, pel. or trit,
30.	PHOSPHORUS,	4	Dilution.
31.	PODOPHYLLIN,	3	Trituration.
32.	PULSATILLA,	3	Dilution or pellets.
33.	RHUS,	3	Dilution.
34.	SANTONIN,	2	Trituration.
35.	Silicea,	6	Trituration.
36.	SPONGIA,	3	Dilution, pel. or trit.
37.	SULPHUR,	4	Trituration, dil. or pel.
38.	TARTARUS EMETICUS,	3	Trituration.
39.	THUJA,	1	Dilution.
40.	VERATRUM,	3	Dilution or pellets.

The twenty-four names in CAPITALS are those of medicines which are most frequently needed, and which will suffice for a family living near a homœopathic pharmacy.

NOTE TO PHARMACISTS.—The attenuations are to be made on the decimal scale. The dilutions should in every case, represent the same quantity of the original substance as the corresponding triturations. The pellets are to be about three millimeters, or one-tenth of an inch, in diameter.

MATERIA MEDICA.

ACONITUM.

NOMENCLATURE.—*Latin name*,—Aconitum Napellus, (Linnæus) ; *Natural Order*, Ranunculaceæ : *English names*,—Monkshood, Wolfsbane, Aconite.

GENERAL DESCRIPTION.—An herbaceous plant, from two to five feet in hight, having palmate leaves and a spiked raceme of blue flowers. One of the sepals arches over and covers the stamens and pistils. All parts of the plant are poisonous. It is a native of Central Europe, but is often cultivated as a garden flower.

PREPARATIONS.—The tincture is made from the entire fresh plant. The preparation for internal use, is the third decimal dilution. A lotion, for immediate external use, is made by adding four parts of water to one part of the tincture.

THERAPEUTIC USES.—The principal use of this remedy is to check local inflammations, in the initial stage. It is of little use after the inflammation is fully established. In the fevers which accompany infectious diseases, such as Scarlet Fever, Measles and Small-pox, it is worse than useless to give Aconite.

CHARACTERISTIC SYMPTOMS.—Mental and nervous excitement; fear; heat and dryness of the skin; scanty urine; quick, full pulse; prickling sensation in the skin.

ÆSCULUS.

NOMENCLATURE.—*Latin name*, Aesculus Hippocastanum, (Linnæus); *Natural Order*, Sapindaceæ; *English name*,—Horse-Chestnut.

GENERAL DESCRIPTION.—A medium-sized tree, bearing digitate leaves composed of seven leaflets, each of which resembles a leaf of the common Chestnut; a large pyramidal panicle of white, purple and yellow spotted flowers; and nuts about an inch and a quarter in diameter. The tree is a native of Asia, but is cultivated for ornament and shade.

PREPARATIONS.—The inner portion of the nut is the part used in medicine. It is prepared in tincture, dilution and trituration. The only preparation mentioned in the section on Therapeutics is the Cerate. This is prepared by boiling the pulverized nut for half an hour in simple cerate or in Cosmoline, and straining through a flannel cloth. The proportion is one to ten.

THERAPEUTIC USES.—In Piles, enlargment of the veins, and congestion of the pelvic vicera.

CHARACTERISTIC SYMPTOMS.—Constipation; dryness of the fæces; dull pain in the rectum; smarting in the anus.

AILANTHUS.

NOMENCLATURE.—*Latin name*, Ailanthus glandulosus, (Desfontaine); *Natural Order*, Simarubaceæ; *English name*, Tree of Heaven.

GENERAL DESCRIPTION.—A large tree, having leaves one and a half to six feet in length, composed of ten to twenty pairs of leaflets and an odd one at the end. It bears a large panicle of small, greenish, ill-scented flowers. It is a native of China, but has been cultivated for shade.

PREPARATIONS.—The tincture is made from the fresh bark. The remedy is generally administered in the first dilution, in doses of one to five drops.

THERAPEUTIC USES.—In Scarlet Fever, Diphtheria, ulcerated sore throat, Dysentery, etc.

CHARACTERISTIC SYMPTOMS.—Delirium ; thin, watery discharge from the nose and throat ; livid appearance of the throat ; dark-red eruption like that of Measles or Scarlet Fever.

APIS.

NOMENCLATURE.—*Latin name*, Apis mellifica, (Linnæus) ; *Order*, Hymenoptera ; *English name*, Honey Bee.

PREPARATIONS.—A trituration of the entire, living worker-bee, in sugar of milk. The third trituration is the one most used.

THERAPEUTIC USES.—In Dropsy, Scarlet Fever, Erysipelas, Nettle-rash, etc.

CHARACTERISTIC SYMPTOMS.—Swelling of the whole body, especially of the feet and the lower eyelids ; redness of the skin.

ARNICA.

NOMENCLATURE.—*Latin name*, Arnica montana, (Linnæus); *Natural Order*, Compositæ ; *English name*, Leopard's-bane.

GENERAL DESCRIPTION.—A perennial, herbaceous plant, about a foot in hight. The leaves are bright green, the lower obtuse and the upper acute. The flowers are large and of a fine orange-color. It grows in Europe.

PREPARATIONS.—The best preparation is a tincture from the entire fresh plant. From this are made the dilutions for internal use. The third dilution is most generally useful. A tincture for external use, is made from the dried flowers. Arnica lotion is prepared extemporaneously from the tincture, by adding six parts of water. Arnica cerate is made by steeping the dried flowers in simple cerate or Cosmoline in the proportion of one to ten, and straining through a cloth.

THERAPEUTIC USES.—Arnica is the best remedy for the effects of falls, and for Bruises and Sprains. It is of great use also in inflammation of the brain.

CHARACTERISTIC SYMPTOMS.—Headache, pain produced by pressing on or moving the injured parts, "black and blue" appearance. It is not always safe to apply Arnica to wounds in which the *skin* is *broken*, for it sometimes produces inflammation of the skin. In these cases, Calendula lotion is preferable.

ARSENICUM.

NOMENCLATURE.—*Latin name*, Arsenicum album, (old authors); *Class*, Mineral Acids; *English name*, White Arsenic; *Synonyms*, Acidum arsenicosum, (later authors), Arsenious acid.

GENERAL DESCRIPTION.—A white or translucent solid, rather freely soluble in water and sparingly so in alcohol. Chemically, it is composed of 75 parts of metallic

Arsenic and 99 parts of Oxygen, by weight. The fatal dose of White Arsenic is from two grains upward.

PREPARATIONS.—The strongest tincture contains one grain of crude Arsenicum in one hundred minims (drops) of the liquid. The third and higher dilutions are prepared from this. The triturations are made, with sugar of milk, from the crude drug. The fourth, fifth and sixth attenuations are most useful.

THERAPEUTIC Uses.—The action of this poison is mainly on the digestive and nutritive functions ; hence its applicability to a very wide range of disordered conditions. Intermittent Fever, Typhoid Fever, Dropsy, Cholera, Diarrhœa and many other diseases present the diseased conditions in which Arsenicum is curative.

CHARACTERISTIC SYMPTOMS.—Constant, intense thirst ; dryness of the lips, mouth and throat ; vomiting ; burning sensation in the throat and stomach ; pale or bluish color of the skin ; cold clammy sweats ; general coldness ; quick, feeble pulse ; bloated condition of the whole body ; diarrhœa ; shortness of breath ; palpitation of the heart ; loss of appetite, flesh and strength ; gangrenous ulcers ; periodical complaints.

BELLADONNA.

NOMENCLATURE.—*Latin name*, Atropa Belladonna, (Linnæus); *Natural Order*, Solanaceæ ; *English name*, Deadly Nightshade.

GENERAL DESCRIPTION.—An herbaceous, perennial plant from three to five feet in hight. The leaves are arranged in pairs, dusky green above, and paler beneath. The flowers are large, bell-shaped, five-parted, solitary

162

in the axils of the leaves and of a pale-purple hue. The ripe berries are black, shining, about as large as cherries, and abound in a purple juice.

PREPARATIONS.—The tincture is made from the entire fresh plant. The third dilution is the one best adapted for general use.

THERAPEUTIC USES.—Inflammation of the throat, Inflammation of the brain, Scarlet Fever and Erysipelas, are the diseases in which Belladonna is most useful.

CHARACTERISTIC SYMPTOMS.—Redness of the face; congestion of the eyes ; beating pain in the forepart of the head; delirium; dilated pupils; convulsions; dryness of the throat; dry, spasmodic cough; inflammatory redness of the skin ; sleeplessness.

BRYONIA.

NOMENCLATURE.—*Latin name*, Bryonia alba ; *Natural Order*, Cucurbitaceæ ; *English name*, White Bryony.

GENERAL DESCRIPTION.—A perennial, herbaceous, climbing plant, having five-lobed, heart shaped leaves and small, yellow flowers. The fruit consists of small, black berries about as large as peas. The plant is a native of Europe.

PREPARATIONS.—The tincture is made from the fresh root. The third dilution is most appropriate for general use.

THERAPEUTIC USES.—The curative virtues of Bryonia are most marked in Rheumatism, Pleurisy and Pneumonia.

CHARACTERISTIC SYMPTOMS.—Headache, worse on stooping ; dry, cracked lips ; bitter taste in the mouth ; yellowness of the skin; stiching pains in the chest; pains worse on motion ; constipation.

CACTUS.

NOMENCLATURE.—*Latin names*, Cereus grandiflorus, (DeCandolle), Cactus grandiflorus, (medical authors); *Natural Order*, Cactaceæ ; *English name*, Night-blooming Cereus.

GENERAL DESCRIPTION.—A creeping, rooting shrub, with a fleshy, prickly stem and no leaves. The flowers are magnificently beautiful and very fragrant. They open at night and endure but a few hours. In size, they measure from eight to twelve inches in diameter. The plant is a native of Mexico and the West Indies.

PREPARATIONS.—The tincture is made from the flowers and stems, in equal parts. The second dilution is the strength in which it has the best therapeutic effects.

THERAPEUTIC USES.—In Heart Diseases and Rheumatism.

CHARACTERISTIC SYMPTOMS.—Palpitation of the heart; dizziness; buzzing noise in the ears; sensation as though there were an iron band about the heart. All of these symptoms are aggravated by exercise.

CALENDULA.

NOMENCLATURE.—*Latin name*, Calendula officinalis, (Linnæus); *Natural Order*, Compositæ; *English name*, Pot Marigold.

GENERAL DESCRIPTION.—An annual herb, about a foot in hight. The leaves are oblong, acute and rough-edged. The flower is large, orange-colored, and solitary at the top of the stem.

PREPARATIONS.—The same as *Arnica*, which see.

12

THERAPEUTIC USES.—A local application in cuts, open wounds and burns. The first dressing may be made with the lotion and the subsequent ones with the cerate, spread on soft linen.

It relieves pain and prevents inflammation, and hastens recovery.

CAMPHORA.

NOMENCLATURE.—*Latin name*, Laurus Camphora, (Linnæus), Camphora officinarum, (Nees); *Natural Order*, Lauraceæ ; *English name*, Camphor-tree.

GENERAL DESCRIPTION.—A large, evergreen tree, growing in eastern Asia. The leaves are ovate-lanceolate in shape, two or three inches in length, bright green above and paler beneath. The flowers are small, white and in little stalked clusters. The fruit is a red berry.

PREPARATIONS.—The gum is extracted by steeping the bark and wood of the tree in boiling water, when it rises to the surface or distils over. It is brought to this country and purified by redistillation. The tincture is prepared by dissolving the gum in pure, strong alcohol.

The varying strength of the tincture of Camphor, has been the cause of several accidents. The following list shows the proportions of camphor-gum and alcohol prescribed by different authorities :

Edinburgh Pharmacopœia...........................one to twelve
Dublin   "   ....one to eight.
United States   "   ....one to eight.
London   "   ....one to six.
British Homœopathic Pharmacopœia.... ....one to six.
German   "   "   ....one to six.

Rubini's Tincture of Camphor contains one-half an ounce of Camphor in one ounce of tincture. The preparation recommended in this work is one part of Camphor to five parts of alcohol. The dose is from one to five drops, on sugar. Water may be added to facilitate swallowing. In Asiatic Cholera, Rubini's tincture may be used in the same doses.

THERAPEUTIC USES.—In Asiatic Cholera, Simple Cholera and Syncope.

CHARACTERISTIC SYMPTOMS.—Sudden attacks of fainting ; great prostration of strength ; cold sweat ; cramps ; profuse watery discharges from the bowels ; blueness of the skin.

CARBO.

NOMENCLATURE.—*Latin name*, Carbo vegetabilis ; *English name*, Charcoal.

GENERAL DESCRIPTION.—A black, insoluble substance, obtained by the partial combustion of wood. It consists mostly of Carbon, but contains also about two per cent of earthy matter. It has the remarkable property of absorbing and holding gases, giving them off again when heated. Hence, its use in purifying water for drinking purposes.

PREPARATIONS.—The preparation used in medicine is made from beech or willow wood. Finely triturated with sugar of milk, it is regarded by some physicians a valuable therapeutic agent, while others think it nearly or quite inert.

CHARACTERISTIC SYMPTOMS.—Rising of gases from the stomach; swelling of the abdomen after meals; offensive breath.

CHAMOMILLA.

NOMENCLATURE.—*Latin name,* Matricaria Chamomilla, (Linnæus); *Natural Order,* Compositæ ; *English name,* German Chamomile.

GENERAL DESCRIPTION.—An annual plant, from one to two feet in hight, and much branched. The leaves are once or twice pinnated, the divisions narrow. The flowers have a rounded, yellow centre and spreading or deflexed, white rays. The plant resembles in appearance the common Mayweed, (Maruta Cotula), but druggists often mistake for it the Anthemis nobilis, which is also called Chamomilla.

PREPARATIONS.—The tincture is prepared from the entire fresh plant, in flower. The third dilution is the one most used in medicine, but the medicine may be given much stronger, without ill effect.

THERAPEUTIC USES.—In complaints from teething, Diarrhœa, and crying of infants.

CHARACTERISTIC SYMPTOMS.—Redness of the face, often of only one side, while the other is pale ; accumulation of wind in the bowels ; greenish colored passages from the bowels.

CHININUM ARSENICOSUM.

NOMENCLATURE.—*Latin name,* Chininum arsenicosum, (in modern Latin, Quiniæ Arsenis); *Class,* Salts; *English name,* Arsenite of Quinia.

GENERAL DESCRIPTION.—This is a chemical combination of Quinia and Arsenious Acid. It is a white, amorphous powder, soluble in alcohol, but not in water.

PREPARATIONS.—The third trituration is most appropriate for ordinary use. In extreme cases, the second trituration may be given in one or two grain doses, every two hours.

THERAPEUTIC USES.— It would naturally be expected that a chemical combination of two of the most powerful tonics and antiperiodics known, would prove a useful remedy. Experience has demonstrated this to be the fact, in regard to this preparation. It is used with success in Intermittent Fever, Dumb Ague, Diphtheria, Malignant Scarlet Fever and other formidable diseases.

CHARACTERISTIC SYMPTOMS.—Intermittent disorders; blood-poisoning; great prostration of the vital powers.

COLCHICUM.

NOMENCLATURE.—*Latin name*, Colchicum autumnale, (Willdenow); *Natural Order*, Melanthaceæ; *English name*, Meadow Saffron.

GENERAL DESCRIPTION.—A curious little plant which grows wild in the meadows of temperate Europe. The leaves appear in spring and disappear in summer, while the flowers appear in autumn. The leaf is grass-like. The flower is trumpet-shaped and six-parted at the top, about six inches in length, one-half concealed under the ground.

PREPARATION.—The tincture is made from the fresh bulb or root. The dose is one drop, every hour or two.

THERAPEUTIC USES.—This is a deservedly popular remedy in Rheumatism and Gout.

CHARACTERISTIC SYMPTOMS.—Pains worse at night; puffy swellings; scanty, high-colored urine.

COSMOLINE.

GENERAL DESCRIPTION.—A peculiar substance obtained from coal oil, by evaporating the volatile constituents of the latter substance. In some properties it resembles wax, in others oil. It melts below blood heat and becomes as thin as water. It is insoluble in water, alcohol, ether or chloroform, but soluble in oil of turpentine. It is soft, bland, and unctuous, and, unlike vegetable and animal oils, never becomes rancid.

THERAPEUTIC USES.—It may be used as a local application, in all cases in which ointments or cerates are applicable; such as, cuts, wounds, burns, sores, and dry eruptions on the skin.

DIOSCOREA.

NOMENCLATURE.—*Latin name,* Dioscorea villosa, (Linnæus); *Natural Order,* Dioscoreaceæ; *English name,* Wild Yam.

GENERAL DESCRIPTION.—A delicate, twining vine, found in thickets in various parts of the United States. The leaves are heart-shaped, pointed, nine to eleven nerved, on long stalks. The male and female flowers are on separate plants. The fruit is a three-winged, membranaceous pod, containing from three to nine seeds. The pod is round in general outline and the prominent wings, about one-third of an inch in width, stand apart at an angle of one hundred and twenty degrees.

PREPARATION.—The tincture is made from the fresh root. The dose is from one to three drops every half hour. It is said that most of the "Wild Yam Root" of the drug-stores is obtained from a species of *Smilax.*

THERAPEUTIC USES.—In Bilious Colic, and Diarrhœa.
CHARACTERISTIC SYMPTOMS.—Crampy pains in the stomach or bowels ; yellow, or dark colored stools.

ERIGERON.

NOMENCLATURE.—*Latin name*, Erigeron Canadense, (Linnæus); *Natural Order*, Compositæ ; *English name*, Canada Fleabane.

GENERAL DESCRIPTION.—A common weed in fields, especially in dry, gravelly soils. The plant is from three to six feet in hight, with a single straight stem, numerous narrow leaves and a panicle of inconspicuous flowers at the top. The ray flowers are forty to fifty in number. The stem is hairy.

PREPARATION.—The tincture is prepared from the entire fresh plant, in flower. The dose is from one to three drops.

THERAPEUTIC USES. — In Hæmorrhage, Dysentery, Diarrhœa, and Gonorrhœa.

CHARACTERISTIC SYMPTOMS.—Flow of bright blood, from any part of the body ; profuse watery diarrhœa; painful urination, with discharge of mucus or blood.

FERRUM.

NOMENCLATURE.—*Latin name*, Ferrum ; *Class*, Metals and their compounds ; *English name*, Iron.

PREPARATION.—There are as many as fifty different chemical combinations of Iron which are used medicinally, such as the pure metal, the Oxides, Acetate, Carbonate, Citrate, Muriate, Nitrate, Phosphate, Sulphate, Bromide and Iodide. For general use I prefer the

Pyro-phosphate. This is chemically, the Sesqui-phos-phate of the Sesqui-oxide of Iron, made soluble in water by the addition of a little Citrate of Ammonia. The dose is three grains of the first trituration, about a quarter of an hour after each meal.

THERAPEUTIC USES.—In Anæmia, Chlorosis and kindred disorders.

CHARACTERISTIC SYMPTOMS.—Paleness of the countenance; gradual loss of flesh without preceptible fever.

GELSEMIUM.

NOMENCLATURE.—*Latin name*, Gelsemium sempervirens, (Aiton); *Natural Order*, Loganiaceæ; *English name*, Yellow Jessamine. The Latin name is spelled by some medical writers,—Gelseminum.

GENERAL DESCRIPTION.—A climbing vine with opposite, lanceolate, short-stalked, shining green leaves. The flowers are bright yellow, funnel-form, an inch to an inch and a half in length, and very fragrant. It grows in low grounds in the south-eastern states. The flowers appear in March and April.

PREPARATION.—The tincture is prepared from the fresh root. The dose for an adult is from one to five drops every hour or two, till an effect is produced. The first dilution may be used with good effect.

THERAPEUTIC USES.—In Intermittent Fever, Neuralgia, Dysuria, Dysmenorrhœa, etc. It should be used only at the beginning of an attack. If it fails to act in twelve hours, some other remedy must be chosen. When the patient "sees double," it is a sign that too much has been given.

CHARACTERISTIC SYMPTOMS.—Dull headache ; chilliness followed by heat and restlessness, with pain in the back ; disorders caused by taking cold ; retention of the urine from spasmodic contraction of the urethra; crampy pain preceding, and at the beginning of, the menstrual flow.

HAMAMELIS.

NOMENCLATURE.—*Latin name,* Hamamelis Virginica, (Linnæus); *Natural Order,* Hamamelaceæ ; *English name,* Witch Hazel.

GENERAL DESCRIPTION.—A shrub, eight to fifteen feet in hight, with a spotted stem, and oval, straight-veined, wavy-toothed leaves, and small yellow flowers. The flowers appear late in autumn as the leaves fall, and continue through the winter. The petals are four in number and strap-shaped. The fruit is a two-celled pod containing two seeds which ripen in the summer. The shrub grows in many parts of the United States.

PREPARATION.—The tincture made from the fresh bark is the most reliable preparation, especially for internal use. The distilled extract may be used externally when the stain of the tincture is objectionable. The dose of the tincture is from one to two drops; that of the distilled extract is from ten to fifteen drops. The lotion for external use is made by adding to the tincture five or six parts of water.

THERAPEUTIC USES.—In Hemorrhage, Piles, and Enlargement of the superficial veins.

CHARACTERISTIC SYMPTOMS.—Flow of dark, or venous blood; external piles; puffiness of the lower limbs, especially at night.

HEPAR SULPHURIS.

NOMENCLATURE.—*Latin name*, Hepar Sulfuris calcareum ; *Class*, Salts ; *English name*, Impure Sulphuret of Calcium. This substance must be distinguished from Hepar Sulphuris kalinum, (Sulphuret of Potassium), mentioned in the chapter on Itch.

GENERAL DESCRIPTION.—A grayish white, amorphous powder, having the odor of rotten eggs. This odor is produced by the Sulph ureted Hydrogen which is evolved in the decomposition of the impurities of the salt.

PREPARATION.—This substance is obtained by heating, in a sealed crucible, a mixture of equal parts of powdered oyster shells and flowers of Sulphur. The sixth decimal trituration is the most reliable preparation.

THERAPEUTIC USES.—Boils ; pimples ; ulcers on the skin ; Milk Crust ; inflammation of the throat.

CHARACTERISTIC SYMPTOMS.—A rapid succession of boils ; cuts heal very slowly ; dry tickling sensation, in the upper and back part of the throat.

HYDRASTIS,

NOMENCLATURE.—*Latin name*, Hydrastis Canadensis, (Linnæus); *Natural Order*, Ranunculaceæ; *English name*, Golden Seal.

GENERAL DESCRIPTION.—A perennial herb, about a foot in hight, having three leaves, one from the root and two from the top of the stem. The leaves are round, heart-shaped at the base, five to seven lobed, and doubly serrated on the edge. When full-grown they measure from four to nine inches in width. The flower appears early in the spring. It is very inconspicuous, having no petals and but three, small, greenish white sepals, which fade as the flower opens. The plant grows in rich woods in various parts of the United States.

PREPARATIONS.—The tincture of the fresh root is the most useful preparation. The dose is one drop. A lotion for immediate local use may be made by adding to the tincture from six to twenty parts of water.

THERAPEUTIC USES.—In Diarrhœa, Dysentery, ulcerations of the mucous membranes, Leucorrhœa, and Gonorrhœa.

CHARACTERISTIC SYMPTOMS.—Passages from the bowels streaked with thick mucus and blood. In ulceration of the mouth and throat, a gargle of the lotion often works a speedy cure.

IGNATIA.

NOMENCLATURE.—*Latin name*, Ignatia amara, (Linnæus); Strychnos Ignatia, (later botanists); *Natural Order*, Apocynaceæ ; *English name*, Saint Ignatius' Bean.

GENERAL DESCRIPTION.—A tree of medium size with very long, slender branches, and opposite, short-stalked, oval, pointed leaves. The flowers are white, tubular and fragrant, clustered in the axils of the leaves. The fruit is about the size of a pear. It consists of a smooth, white, woody rind, enclosing about twenty seeds. The seeds are hard, brown and irregular in shape. The tree is found on the south-east coast of Asia.

PREPARATION.—The tincture is made from the seeds. The third dilution is the one which is most used in medicine.

THERAPEUTIC USES.—In nervous prostration, Hysteria, Hypochondriasis, Chorea and Epilepsy.

CHARACTERISTIC SYMPTOMS.—Sensation as though there were a ball in the stomach ; distress in the region of the heart; constipation ; soreness in the spinal column.

IPECACUANHA.

NOMENCLATURE.—*Latin name*, Cephaelis Ipecacuanha, (Martius); *Natural Order* Rubiaceæ ; *English name*, Ipecacuanha.

GENERAL DESCRIPTION.—A small, shrubby plant, about a foot in hight, having from three to six leaves near the top. The leaves are acute, three or four inches long and about half as wide. The flowers are small and white, in a cluster of eight or ten. The fruit is a black berry containg two seeds. The root is from four to six inches long, and about a quarter of an inch in thickness and is marked by peculiar ring-like ridges. The plant is found in Brazil and other parts of tropical South America.

PREPARATION.—The tincture is made from the root. The third dilution is most useful in medicine.

THERAPEUTIC USES.—In Dyspepsia, indigestion, diarrhœa, Whooping-cough, Asthma, etc.

CHARACTERISTIC SYMPTOMS.—Constant nausea ; frequent retching and vomiting ; diarrhœa with stools like yeast ; dry, spasmodic cough ; spitting of blood ; moist, yellowish-coated tongue ; hemorrhage from the uterus.

IRIS.

NOMENCLATURE —*Latin name*, Iris versicolor, (Linnæus); *Natural Order*, Iridaceæ; *English name*, Larger Blue Flag.

GENERAL DESCRIPTION.—Stem about three feet high ; leaves sword-shaped, three-fourths of an inch wide and about as long as the stem ; flower two and a half to three inches long, funnel-form, six-parted, the three sepals curved backward, the three petals straight, sepals and petals blue and white ; fruit a three-angled pod, two to three inches in length, and three-fourths of an inch in thickness   The plant is common in wet places, throughout the United States.

PREPARATIONS.—The tincture is made from the fresh root.   The second and third dilutions are the best for general use.   In some cases the mother tincture does better.

THERAPEUTIC USES.—In Diarrhœa, Sick Headache, Milk Crust, etc.

CHARACTERISTIC SYMPTOMS.—Pain in one side of the head ; vomiting of sour, bilious matters ; watery diarrhœa, with griping pain in the bowels.

KALI BICHROMICUM.

NOMENCLATURE.–*Latin name*, Kali chromicum rubrum; *Class*, Salts ; *English name*, Bichromate of Potassium.

GENERAL DESCRIPTION.—This substance exists in the form of orange-red, translucent crystals of considerable size.   It is soluble in water but not in alcohol.   Chemically, it consists of two equivalents of Chromic Acid and one of Potassium.

PREPARATION.—The third trituration is most convenient, if the medicine is to be taken directly into the
mouth. The second trituration may be taken, dissolved
in water, in such strength as to give a slight tinge of color
to the water. The dose is two grains of the third trituration, or a teaspoonful of the watery solution.

THERAPEUTIC USES.—This is a valuable remedy in
Catarrh, Bronchitis, Croup and Diphtheria.

CHARACTERISTIC SYMPTOMS.—Secretion of a tough,
stringy, adhesive mucus ; expectoration streaked with
blood ; yellowish coat on the tongue.

KALI BROMATUM.

NOMENCLATURE.—*Latin name*, Kali bromatum ; *Class*,
Haloid Salts ; *English name*, Bromide of Potassium.

GENERAL DESCRIPTION.—This salt crystallizes in the
form of transparent, rectangular prisms. It is freely soluble in water, but very slightly soluble in alcohol. Its taste
is similar to that of common salt but is more pungent.
Chemically, it consists of one equivalent of Potassium
and one of Bromine.

PREPARATION.—The therapeutic effects of this substance are obtained from the crude salt. It may be
given, dissolved in water, or if the taste is objectionable,
in milk. The dose for an adult is from ten to twenty
grains every six hours. A child may take half a grain
every hour till an effect is produced.

THERAPEUTIC USES.—In Whooping-Cough, irritation
of teething, Epilepsy, Chorea, Meningitis, and Lead-
poisoning.

CHARACTERISTIC SYMPTOMS.—Nervousness ; sleepless-ness; pain in the back of the head and neck; simple spas-modic cough. In Epilepsy and Chorea the remedy must be given daily, in full doses, for some months. In Whooping-Cough the dose may be repeated according to the frequency and severity of the paroxysms. In irrita-tion of the brain from teething, a strong solution in water may be rubbed on the gums every twenty minutes.

KALI IODATUM.

NOMENCLATURE.—*Latin name*, Kali iodatum ; *Class*, Haloid Salts; *English name*, Iodide of Potassium. Some medical authors have incorrectly called this salt " Kali hydriodicum." It contains no hydrogen.

GENERAL DESCRIPTION.—Iodide of Potassium crys-talizes in beautiful transparent cubes. It is soluble in both alcohol and water. It has an unpleasant, salty taste.

PREPARATION.—The second decimal dilution, in alco-hol is the best form for preservation and administration. This contains one grain of the salt in one hundred minims of the solution.

THERAPEUTIC USES.—This remedy is serviceable in Catarrh, Asthma, Rheumatism, Gout, Syphilis, etc.

CHARACTERISTIC SYMPTOMS.—Discharge of a watery fluid from the nose and eyes; stuffed feeling in the upper part of the chest ; aching pains in the bones, worse at night ; enlarged glands ; symptoms of poisoning by mercury ; swellings on the bones.

MACROTIN OR CIMICIFUGIN.

NOMENCLATURE.—*Latin name*, Cimicifuga racemosa, (Elliott); *Natural Order*, Ranunculaceæ ; *English name*, Black Snakeroot ; *Synonyms*, Macrotys racemosa, (Rafinesque); Actæa racemosa, (Linnæus).

GENERAL DESCRIPTION.—The plant from which Macrotin is obtained, is a perennial herb, from four to eight feet in hight. The leaves are very large, dividing and subdividing into threes. The leaflets are oblong in shape and irregularly saw-toothed on the edges. The flowers are small and white, and grow in branching, pyramidal racemes, which are from one to three feet in length. The fruit is a dry, oval pod. The plant grows in woods throughout the eastern portion of the United States.

PREPARATIONS.—Macrotin is fine, dark brown powder, manufactured from the fluid extract of the root. This is usually administered in the third trituration. The tincture of the plant, called *Cimicifuga*, is equally good if not better, except that the powder is more pleasant and convenient to take. The dose of the third trituration is two grains ; of the tincture, one drop.

THERAPEUTIC USES.—In disordered menstruation, Leucorrhœa, Hysteria, Rheumatism, etc.

CHARACTERISTIC SYMPTOMS.—Pain in the back and loins ; stiffness of the neck and back ; dragging sensation in the lower part of the bowels ; suppression or delay of the menses ; pain in the top of the head.

MERCURIUS CORROSIVUS.

NOMENCLATURE.—*Latin name*, Mercurius corrosivus sublimatus; *Class*, Haloid Salts ; *English name*, Corrosive Chloride of Mercury.

GENERAL DESCRIPTION.—This salt exists in colorless crystals, soluble in water and alcohol. In the crude state, it is extremely poisonous. Chemically, it contains by weight, 70 parts of Chlorine and 200 parts of Mercury. Some chemists have called it Proto-chloride and others Bi-chloride of Mercury.

PREPARATION.—The third and fourth attenuations are the most useful in medicine. It may be prepared in alcohol or in sugar of milk.

THERAPEUTIC USES.—In Dysentery, Chronic Catarrh, inflammation of the eyes, Gonorrhœa, Syphilis, etc.

CHARACTERISTIC SYMPTOMS.--Thick, yellowish secretion from the mucous membrane ; passages of blood and mucus from the bowels ; straining in the rectum after the passages.

For Inflammation of the eyes, a lotion made by dissolving one part of the third attenuation in ten parts of water, forms an excellent remedy. For Chronic Catarrh, the third trituration snuffed vigorously into the nostrils, once a day, has been, in my experience, the most successful treatment of any yet discovered. In Dysentery, a good rule for administration is to give a dose immediately after each movement of the bowels.

13

MERCURIUS IODATUS RUBER.

NOMENCLATURE.—*Latin name*, Mercurius iodatus ruber, *Class*, Haloid Salts ; *English name*, Red Iodide of Mercury. This is also called by medical writers, Biniodide of Mercury, (*Latin*, Mercurius biniodatus), but the same name has also been applied by good chemists to the Yellowish-Green Iodide of Mercury, (*Latin*, Mercurius iodatus flavus).

GENERAL DESCRIPTION.—This substance exists in crystals of bright vermilion color, which turn yellow when heated and gradually become red again on cooling. It is somewhat soluble in alcohol but almost insoluble in water. Chemically, it contains, by weight, 100 parts of Mercury and 127 parts of Iodine.

PREPARATIONS.—The tincture contains one grain of the crystals in two hundred minims of alcohol. The third dilution is prepared by adding to the tincture, four times its bulk of alcohol. The dose of the third dilution is from one to five drops. The third trituration of the crystals may be taken dry; but does not dissolve when thrown into a glass of water. The dose is one or two grains.

THERAPEUTIC USES-—In Quinsy, Diphtheria, Goitre enlargment of the lymphatic glands, chronic congestion of the liver, etc.

CHARACTERISTIC SYMPTOMS.—Hardness and swelling of the glands ; yellowness of the skin, and of the whites of the eyes ; constipation.

MERCURIUS VIVUS.

NOMENCLATURE,—*Latin name*, Mercurius ; *Class*, Elements; *English names*, Metallic Mercury, and Quicksilver.

GENERAL DESCRIPTION.—Pure Mercury is a brilliant, heavy, odorless and tasteless liquid. In nature, it is found sometimes in the pure state, but more frequently in combination with Sulphur or with Silver.

PREPARATION.—The third trituration is most useful in practice.

THERAPEUTIC USES.—In inflammation of the throat, glandular swellings, Diarrhœa and Syphilis.

The curative effects of this powerful agent, have, in years past, been greatly over-rated, and, as a consequence, the remedy has fallen into ill-repute. Many lives have been destroyed, and the health of many has been undermined, by poisonous doses of Mercury, administered as a medicine. Owing to its easy divisibility, it is readily introduced into the circulation, but owing to its insolubility, it is excreted very slowly. It has been found in the metallic state, in the brain, in the lungs, in the liver, in the bones and in the cellular tissues. It may be introduced through the skin, by rubbing it on in the form of ointment. In view of these facts, I prefer to give, and recommend for use, only the soluble salts of Mercury, which are excreted as readily as they are absorbed. The M. corrosivus and M. iodatus ruber are soluble. The Mercurius solubilis Hahnemanni is not soluble, as its name would seem to imply. For this reason, its use should be limited if it is given at all.

CHARACTERISTIC SYMPTOMS.—The deposits of Secondary Syphilis seem to soften and disappear more rapidly under the action of this, than with any other remedy. Most, if not all, of the other indications for Mercurius are met by the M. iodatus or the M. corrosivus.

NITRI ACIDUM.

NOMENCLATURE.—*Latin name*, Nitri Acidum ; (*modern Latin*, Acidum nitricum); *Class*, Acids ; *English name*, Nitric Acid.

GENERAL DESCRIPTION.—Pure Nitric Acid is a sour, heavy, colorless liquid, which is capable of corroding all animal and vegetable as well as many mineral substances. It contains, by weight, Oxygen 144 parts, Nitrogen 28 parts, and Hydrogen 8 parts.

PREPARATIONS.—The Nitric Acid of the United States and the London Pharmacopœias, (specific gravity 1.42), contains, by weight, in five parts, three parts of the pure Acid, (specific gravity 1.50). The first decimal dilution is prepared by adding to one part of the Nitric Acid, U. S., five parts of distilled water. The subsequent dilutions must be also made in distilled water, and, like the stronger preparations, kept only in glass-stoppered vials. The first dilution is the most certain in its effects. The dose is from one to five drops, in water. Never use a spoon in handling the acid.

THERAPEUTIC USES.—Internally, in Consumption Typhoid Fever, Syphilis, Mercurial poisoning and other wasting diseases ; locally, as a caustic in poisoned wounds, and syphilitic ulcers.

CHARACTERISTIC SYMPTOMS.—Profuse sweating; chronic or wasting diarrhœa ; sponginess of the gums; eating ulcers.

NUX VOMICA.

NOMENCLATURE.—*Latin name*, Strychnos Nux vomica, (Linnæus); *Natural Order*, Apocynaceæ ; *English name*, Dog Button.

GENERAL DESCRIPTION.—The tree is of medium size, with long, drooping branches and opposite, roundish, oval leaves. The flowers are small, white, funnel-shaped, in flat clusters at the ends of the branches. The fruit is about as large as an orange. It has a hard, fragile shell and a soft, juicy pulp, enclosing numerous seeds. The seeds are hard, circular, flat, about five-eighths of an inch in diameter and one-eighth of an inch in thickness.

PREPARATION.—The seeds are the part used in medicine. The third dilution, made from the tincture and the third trituration, made from the dry seeds, are the most useful preparations.

THERAPEUTIC USES.—In Dyspepsia, Colic, Constipation, Chorea, and for the effects of stimulants or sedentary habits.

CHARACTERISTIC SYMPTOMS.—Dull headache with dizziness ; accumulation of wind in the bowels ; sour stomach; constipation with frequent desire for a passage; short, dry, spasmodic cough; nervousness ; sexual debility.

PHOSPHORUS.

General Description.—The element Phosphorus, in its most common form, is a semi-transparent, waxy body which takes fire at 100° F. or, if air is excluded, melts at 108° F. and boils at 550° F. It rapidly changes to Phosphoric Acid, when exposed to the air, at ordinary temperatures.

Preparations.—The alcoholic tincture contains one part of pure Phosphorus, in seven hundred and fifty parts of the solution. The fourth dilution is the one most applicable to general medicinal use. The pills and powders are very unreliable, and even the dilutions should not be kept longer than a year.

Therapeutic Uses.—In inflammation of the lungs, inflammation of the liver, and nervous debility.

Characteristic Symptoms.—Short, dry, suffocative cough, with scanty, frothy expectoration ; yellowness of the skin ; pain in the region of the liver ; debility from prolonged nervous excitement.

PODOPHYLLUM.

Nomenclature.—*Latin name*, Podophyllum peltatum, (Linnæus); *Natural Order*, Berberidaceæ; *English names*, Mandrake and May-Apple.

General Description.—A perennial, herbaceous plant, about a foot and a half in hight, found growing in woods throughout the United States. The leaves are of two kinds. Those of the flowerless stems are from four to nine lobed, with the stem in the centre. Those of the flowering stem are similarly lobed, two in number, having the stem on the side. The leaves are from six to twelve inches in diameter, and resemble an open umbrella. The flowers are on a stalk which grows from the point where the two leaves branch out. They are white, fragrant and nodding, about two inches in diameter. The petals are obtuse and from six to nine in number. The fruit is round-oval in shape, about an inch in diameter, and, when ripe, pleasant to the taste, and harmless if eaten in small quantities.

PREPARATIONS.—The tincture is made from the fresh root. The dose is one drop.

Podophyllin is a yellow powder made from the tincture or fluid extract. This is given in the second and third triturations, in doses of two or three grains. It is more convenient than the tincture and probably as efficacious.

THERAPEUTIC USES.—In Biliousness, Constipation, and Diarrhœa.

CHARACTERISTIC SYMPTOMS.—Coated tongue ; dull, dizzy headache ; constipation, with occasional attacks of diarrhœa : yellowish hue of the skin.

PULSATILLA.

NOMENCLATURE.—*Latin names*, Pulsatilla nigricans and Anemone pratensis ; *Natural Order*, Ranunculaceæ ; *English names*, Pasque flower, Wind flower, and Meadow Anemone.

GENERAL DESCRIPTION.—An herbaceous plant with a perennial root, found growing in dry pastures in middle Europe. The leaf stalks grow directly from the root; the leaves are bi-pinnate, and the segments short and narrow. The stem in fruit is from five to eight inches in hight.

The flowers appear early in the spring. The petals are wanting, but the sepals are conspicuous. The latter are of a dark purple color. The seeds have long, feathery plumes which enable them to be transported by the wind.

PREPARATION.—The tincture is made from the entire fresh plant in flower. The third dilution is preferred for ordinary practice.

THERAPEUTIC USES.—In acute inflammation of the eyes or ears, Measles, Mumps, Dyspepsia, irregularity of the menses, etc.

CHARACTERISTIC SYMPTOMS.—Watery discharge from the eyes, nose or ears ; earache ; transitory pains in the stomach and various parts ; pain in the side, in front of the hip bone; delayed menstruation; swelling of the testicles ; indigestion from eating fatty food.

This remedy performs most cures in persons of light hair, blue eyes and mild, tearful disposition.

RHUS.

NOMENCLATURE.—*Latin name*, Rhus Toxicodendron ; *Natural Order*, Anacardiaceae; *English names*, Poison Ivy and Poison Oak.

GENERAL DESCRIPTION.—A perennial plant with a woody stem from one to three feet in hight, and rooting runners from three to sixty feet in length. The leaves are composed of three, bright green, oval, pointed leaflets which are entire or notched on the edge. The flowers are five-parted, greenish and inconspicuous· The fruit consists of dry berries about an eighth of an inch in diameter. The whole plant has a poisonous milky juice which, in drying, loses its poisonous properties and forms an indelible black ink. When the runners climb fences and trees, the plant is called Rhus radicans. It is a common pest in meadows, in the United States, growing near fences, trees or stone-heaps.

If the juice is brought in contact with the skin, it produces a burning, itching, vesicular eruption, and an inflammation of the skin resembling Erysipelas, which lasts about two weeks and terminates in a peeling off of the epidermis. (Common soft soap is one of the best local applications for the relief of the troublesome itching of the early stage. A lotion consisting of four parts of lime water, one part tincture of Gelsemium and one part glycerine will relieve the swelling and shorten the course of the inflammation.)

THERAPEUTIC USES.—In Erysipelas, Nettle-rash, Rheumatism, Sprains, etc.

CHARACTERISTIC SYMPTOMS.—Burning itching of the skin with inflammatory redness and the formation of small, watery vesicles. Painful stiffness of the joints, worse on first moving but relieved by continued movement.

SANTONIN.

GENERAL DESCRIPTION.—A white, crystalline substance obtained from the dried flowers of the European Wormseed (Artemisia contra).

PREPARATIONS.—The first and second triturations.

THERAPEUTIC USES.—For Worms and irritation of the brain from indigestion.

CHARACTERISTIC SYMPTOMS.—Grinding of the teeth ; paleness of the face ; itching of the nose ; spasmodic, hollow cough ; starting in sleep ; convulsions

SILICEA.

NOMENCLATURE.—*Latin name*, Silex ; *Class*, Minerals ; *English name*, Quartz ; *Chemical name*, Silica or Oxide of Silicon.

GENERAL DESCRIPTION.—This substance occurs in nature in the form of six-angled, transparent prisms or pyramids of very great hardness.

PREPARATION.—The sixth decimal trituration gives the best curative results.

THERAPEUTIC USES.—In Ulcers, Felons, scrofulous Abscesses, and chronic diseases of the bones.

CHARACTERISTIC SYMPTOMS.—Prolonged suppuration ; throbbing pain in the fibrous tissues ; fœtid sweat of the feet.

SPONGIA.

NOMENCLATURE—*Latin name*, Spongia; *Class*, Rhizopods ; *English name*, Sea Sponge.

GENERAL DESCRIPTION.—The living sponge consists of a colony of sarcoids, supported by a framework of horny, calcareous or silicious material. The individual sarcoid is a minute, jelly-like animal, which has a cilium, or hair-like appendage capable of vibrating in the water with great rapidity. The framework of the sponge is secreted from the bodies of the sarcoids like the shell of a snail or oyster. The whole colony of sarcoids, in the aggregate, forms a slimy mass which covers the outside of the sponge and lines its cavities. By the motion of the cilia or appendages, a current of water is drawn in at the smaller openings and forced out at the larger openings or canals of the sponge, thus supplying the whole colony with nutriment.

Before the sponges are brought to market, they are dried in the sand and then exposed to the action of the tides to be washed. The best sponges are those which contain the largest proportion of horn and the smallest proportion of lime and silica. These are brought from Smyrna, on the Mediterranean Sea.

PREPARATION.—The sponge is prepared for use in medicine by roasting in a closed vessel. The third trituration and the third dilution are the proper forms for dispensing.

THERAPEUTIC USES.—In Croup, Whooping-cough, and Goitre.

CHARACTERISTIC SYMPTOMS.—Dryness of the throat and wind-pipe; hoarse, dry, hollow cough; swelling of the glands in scrofulous children.

SULPHUR.

NOMENCLATURE.—Latin name, Sulfur; Class, Elements; English name, Brimstone.

GENERAL DESCRIPTION.—A pale yellow, crystalline substance which melts at 240° F. and takes fire at 300° F. It is insoluble in water and very slightly soluble in alcohol. It is prepared for medicinal use by distilling and washing in water.

PREPARATIONS.—The saturated alcoholic tincture, at ordinary temperatures, contains about one part of Sulphur in two thousand parts of tincture; hence, the fourth dilution is prepared by adding to the tincture four times its bulk of alcohol. The fourth trituration is however a more uniform and reliable preparation.

THERAPEUTIC USES.—In chronic skin diseases, eruptions, Constipation, Piles, and irritation of the bladder

CHARACTERISTIC SYMPTOMS.—Itching eruptions; heat in the hands and face while the rest of the body is cold; frequent, scalding urination.

TARTARUS EMETICUS.

NOMENCLATURE.—*Latin name,* Tartarus emeticus; *Modern Latin,* Antimonii et Potassii Tartras; *Class,* Salts; *English names,* Tartar emetic, Tartrate of Antimony and Potassium. It is sometimes called Antimonium tartaricum or Tartarized Antimony.

GENERAL DESCRIPTION.—This salt is obtained in the form of transparent crystals of a nauseous taste. It is soluble in water and sparingly so in dilute alcohol.

PREPARATION.—The third trituration is the most appropriate form for general use. The dose for an adult is two grains. The quantity may be reduced for infants and children, by dissolving five grains in twenty teaspoonfuls of water and giving teaspoonful doses of the solution.

THERAEUTIC USES..—In Catarrh, Bronchitis, Asthma, Pneumonia, Diarrhœa and Small-Pox.

CHARACTERISTIC SYMPTOMS.—Accumulation of loose mucus in the windpipe and bronchial tubes ; nausea ; watery diarrhœa ; pustular eruptions on the skin.

THUJA.

NOMENCLATURE.—*Latin name*, Thuja occidentalis ; (Linnæus); *Natural Order*, Coniferæ ; *English name*, Arbor Vitæ, (Tree of Life).

GENERAL DESCRIPTION.—A large tree, growing in the northern United States, on the rocky banks of lakes and rivers and occasionally in swamps. The leaves are evergreen, from one-sixteenth to one-fourth of an inch in length, in four rows, pressed close to the young branches. On the back of each leaf is a little sac containing a transparent, pungent, aromatic oil. The branches are much flattened and spread out horizontally. The flowers grow in a compact, ovoid cone about a fourth of an inch in length. The fruit is a small two-winged seed. There are from ten to fifteen in each cone.

PREPARATIONS.—The tincture is made from the fresh leaves and young shoots. The first dilution is the one recommended for use.

THERAPEUTIC USES.—Inflammation or irritation of the bladder or urethra ; Warts.

CHARACTERISTIC SYMPTOMS.—Burning sensation at the end of the urethra ; chronic discharge of mucus and pus from the urethra.

VERATRUM.

NOMENCLATURE.—*Latin name*, Veratrum album, (Willdenow); *Natural Order*, Melanthaceæ; *English name*, White Hellebore.

GENERAL DESCRIPTION.—A perennial, herbaceous plant from two to four feet in hight, found growing in Alpine pastures in Europe. The leaves are thick and strongly nerved and plaited from end to end. The flowers are greenish and six-parted, in a large panicle, at the top of the strong, leafy stem.

PREPARATION.—The tincture is made from the dried root. The third dilution is most useful in practice.

THERAPEUTIC USES.—In Colic, Diarrhœa, Cholera, and Spasmodic Asthma.

CHARACTERISTIC SYMPTOMS.—Watery diarrhœa, accompanied by severe crampy pain in the bowels ; vomiting ; cold sweat and feeble pulse.

INDEX.

14

www.ingramcontent.com/pod-product-compliance
Lightning Source LLC
Chambersburg PA
CBHW030832270326
41923CB00007B/1021